Realm of
ALGEBRA

Isaac Asimov

diagrams by Robert Belmore

HOUGHTON MIFFLIN COMPANY BOSTON
The Riverside Press Cambridge

Some other books
by Isaac Asimov

Words of Science
Realm of Numbers
Breakthroughs in Science
Realm of Measure
Words from the Myths

FIFTH PRINTING   R

CONTENTS

To

Robert P. Mills,

Kindly Editor

1

The Mighty Question Mark

ALGEBRA IS just a variety of arithmetic.

Does that startle you? Do you find it hard to believe? Perhaps so, because the way most of us go through our schooling, arithmetic seems an "easy" subject taught in the lower grades, and algebra is a "hard" subject taught in the higher grades. What's more, arithmetic deals with good, honest numbers, while algebra seems to be made up of all sorts of confusing x's and y's.

But I still say there's practically no difference between them and I will try to prove that to you.

Let's begin by saying that if you had six apples and I gave you five more apples, you would have eleven apples. If you had six books and I gave you five more books, you would have eleven books. If you had six dandelions and I gave you five more dandelions, you would have eleven dandelions.

I don't have to go on that way, do I? You can see that if you had six of any sort of thing at all and I gave you five more of that same thing, you

would end with eleven of it altogether. So we can
forget about the actual object we're dealing with,
whether apples, books, dandelions, or anything else,
and just concentrate on the numbers. We can say
simply that six and five are eleven, or that six plus
five equals eleven.

Now people are always dealing with numbers;
whether in the work they do, in the hobbies they
pursue, or in the games they play. They must
always remember, or be able to figure out if they
don't remember, that six plus five equals eleven, or
that twenty-six plus fifty-eight equals eighty-four,
and so on. What's more, they often have to write
down such arithmetical statements. But the writing
can get tedious, particularly where the numbers
grow large and complicated.

For that reason, ever since the earliest days of
civilization, people have been trying to figure out
good short-cuts for writing down numbers. The
best system ever invented was developed in India
some time in the 800's. In that system, each
number from one to nine had its own special mark.
The marks we use these days in our country are
1, 2, 3, 4, 5, 6, 7, 8, and 9. In addition, the system
includes a mark for zero, which we write as 0.

Any mark written down as a short-cut method
of representing something is called a "symbol."
(The very words you are now reading are symbols

of the various sounds we make when we speak, and the sound we make when we say "horse" is just a symbol of the actual creature itself.)

The marks I have written two paragraphs ago are symbols for the first nine numbers and zero, and may be called "numerical symbols." The Arabs picked them up from the mathematicians of India and passed them on to the Europeans in about the tenth century. We still call these numerical symbols the "Arabic numerals," in consequence.

All numbers higher than nine can be written by using combinations of these numerical symbols according to a system which I won't explain here because it is so familiar to you.* Thus, the number twenty-three is written 23, while seven hundred and fifty-two is written 752.

You can see how handy numerical symbols can be. In fact, they are so handy that you would never see anyone write: "The total is six thousand seven hundred and fifty-two." It would always be written, "The total is 6752." A great deal of space and effort is saved by writing the numerical symbols in place of words, yet you are so accustomed to the

---

* Actually, I have explained the number system in a book I wrote called *Realm of Numbers*, published in 1959 by Houghton Mifflin Company. You don't have to read it to understand this book, but you might find it useful in explaining some arithmetical points I will have to skip over a little quickly here.

symbols that you read a sentence containing them
just as though the words had been spelled out.

Nor are the numerals the only symbols used in
everyday affairs. In business, it is so usual to have
to deal with dollars that a symbol is used to save
time and space. It is $, which is called the "dollar
sign." People just read it automatically as though
it were the word itself so that $7 is always read
"seven dollars." There is also ¢ for "cents," % for
"per cent," & for "and," and so on.

So you see you are completely at home with
symbols.

There's no reason why we can't use symbols to
express almost anything we wish. For instance, in
the statement six plus five equals eleven, we can
replace six by 6, five by 5, and eleven by 11, but
we don't have to stop there. We can have a symbol
for "plus" and one for "equals." The symbol for
"plus" is + and the symbol for "equals" is =.

We therefore write the statement: $6 + 5 = 11$.

FACING THE UNKNOWN

We are so familiar with these symbols and with
others, such as − for subtraction, × for multiplication, and ÷ for division that we give them no
thought. We learn them early in school and they're
with us for life.

But then, later in our schooling, when we pick up

new symbols, we are sometimes uneasy about them
because they seem strange and unnatural, not like
the ones we learned as small children. Yet why
shouldn't we learn new symbols to express new
ideas? And why should we hesitate to treat the
new symbols as boldly and as fearlessly as we treat
the old?

Let me show you what I mean. When we first
start learning arithmetic, what we need most of all
is practice, so that we will get used to handling
numbers. Consequently, we are constantly presented with numerous questions such as: How
much is two and two? If you take five from eight,
how much do you have left?

To write these questions down, it is natural to
use symbols. Therefore, on your paper or on the
blackboard will be written

$$2 + 2 =$$
$$8 - 5 =$$

and you will have to fill in the answers, which, of
course, are 4 and 3 respectively.

But there's one symbol missing. What you are
really saying is: "Two plus two equals *what?*";
"Eight minus five equals *what?*"

Well, you have good symbols for "two," "eight,"
"five," "plus," "minus," and "equals," but you
don't have a symbol for "what?" Why not have

one? Since we are asking a question, we might use a question mark for the purpose. Then, we can write

$$2 + 2 = ?$$
$$8 - 5 = ?$$

The ? is a new symbol that you are not used to and that might make you uneasy just for that reason. However, it is merely a symbol representing something. It represents an "unknown." You always know just what 2 means. It always stands for "two." In the same way + always stands for "plus." The symbol ?, as I've used it here, however, can stand for any number. In the first case, it stands for 4; in the second case, it stands for 3. You can't know what it stands for, in any particular case, unless you work out the arithmetical problem.

Of course, in the cases given above, you can see the answer at a glance. You can't, though, in more complicated problems. In the problem

$$3312 \div 46 = ?$$

? equals a particular number, but you can't tell which one until you work out the division. (I won't keep you in suspense because this is not a book of problems. In this case, ? stands for 72.)

But you might wonder if I'm just making something out of nothing. Why put in the question mark after all? Why not leave the space blank and just fill in the answer as, in fact, is usually done? Well, the purpose of symbols is to make life simpler. The eye tends to skip over a blank space, and you have no way of reading a blank space. You want to fill the space with a mark of some sort just to show that something belongs there, even if you don't know exactly what for a moment.

Suppose, for instance, you had a number of apples, but weren't sure exactly how many. However, a friend gave you five apples and, after that gift, you counted your apples and found you had eight altogether. How many did you have to begin with?

What this boils down to is that some number plus five equals eight. You don't know what that "some number" is until you think about it a little. The "some number" is an unknown. So you can write

$$? + 5 = 8$$

and read that as, "What number plus five equals eight?" If you had tried to do away with symbols such as a question mark and just left a blank space, you would have had to write

$$+5 = 8$$

and you will admit that that looks funny and is

hard to read. So the question mark, you see, comes in handy.

Now would you be surprised to know that we are already talking algebra? Well, we are. As soon as we begin using a symbol for an unknown quantity, we are in the realm of algebra. Most arithmetic books, even in the very early grades, start using question marks as I have been doing, and they're teaching algebra when they do so.

But this is just arithmetic, you may be thinking.

Exactly! And that is what I said at the very start. Algebra *is* arithmetic, only broader and better, as you will see as you continue reading the book.

It teaches a way of handling symbols that is so useful in considering the world about us that all of modern science is based on it. Scientists couldn't discuss the things that go on about us unless they could use symbols. And even after they've done that, they couldn't handle the symbols properly unless they knew the rules that will be worked out in this book.

INTRODUCING THE LETTER

Actually, the question mark is not a very good symbol for an unknown. It's hard to read because we usually come across it at the end of a question where we don't read it. And if we force ourselves to read it, the result is a three-syllable phrase: "question mark."

The natural symbol to use would be some letter because everyone is familiar with letters and is used to reading them. The trouble with that is, however, that until Arabic numerals were introduced, people used letters as numerical symbols. The Roman system was to use V for "five," X for "ten," D for "five hundred," and so on. If you tried to use letters to represent unknown values as well, there would be endless confusion.

Once the Arabic numerals came in, however, that freed the letters for other uses. Even so, it took centuries for mathematicians to think of using the letters. (Believe it or not, it is very hard to think of good symbols. And often the lack of a good symbol can delay progress in human thought for centuries. A little thing like writing 0 for "zero" revolutionized mathematics, for instance.)

The first person to use letters as symbols for unknowns was a French mathematician named François Vieta (fran-SWAH vee-AY-ta). He did this about 1590 and is sometimes called "the father of algebra" because of it.

Of course, there is still the chance of confusing letters that stand for unknown quantities with letters that form parts of words. For this reason, it quickly became customary to use the letters at the end of the alphabet to symbolize unknowns. Those letters were least frequently used in ordinary words, so there would be least chance for confusion. The least-used letter of all is x, so that is used most commonly to symbolize an unknown.

To allow even less chance of confusion, I will write x and any other such symbol for an unknown in italics throughout this book. Thus, when the letter is part of a word it would be "x"; when it is a symbol for the unknown it will be x.

Now, instead of writing "? $+ 5 = 8$," we would write "$x + 5 = 8$."

Do you see what an improvement this is? First of all, x is a familiar symbol, which we are used to reading and which can be said in one syllable, "eks."

Of course, as soon as some people see the x they feel frightened. It begins to look like algebra. But it's just a symbol doing the same job as the ? that is to be found in all elementary arithmetic books. It happens to be a "literal symbol" (one consisting of a letter) instead of a numerical symbol, but the same rules apply to both. If you can handle the symbol 4, you can handle the symbol x.

2

Setting Things Equal

THE IMPORTANCE OF =

Now THAT we have x, let's find out how to handle it. Since I said, at the end of the last chapter, that it could be handled in the same way ordinary numbers are, let's start with ordinary numbers.

Consider the expression

$$3 + 5 = 8$$

Notice, first, that it has an "equals sign" in it. There are symbols to the left of the "equals sign" and symbols to the right of it, and both sets of symbols, left and right, represent the same quantity. The symbol to the right, 8, represents "eight." The symbols to the left, "$3 + 5$," represent "three plus five," and that comes out to "eight" also.

Whenever you have an "equals sign" with symbols on both sides, each set of symbols representing the same quantity, you have an "equation." (This word comes from a Latin word meaning "to set equal.") The word "equation" may make you think of "hard" mathematics, but you can see that as soon as the school child works out the simplest sum, an equation is involved.

Of course, in order for a set of symbols to make up an equation, they must represent equal quantities on both sides of the equals sign. The expression, $4 + 5 = 8$, is *not* a true equation; it is a false one. In mathematics, naturally, we try to deal with true equations only.

So let's switch our attention now to an equation which has a literal symbol in it, as is the case with

$$x + 5 = 8$$

The symbol x can represent any number, to be sure, but when it is part of an expression containing an equals sign, it is only reasonable to want it to express only those numbers that make a true equation out of the expression. If, in the expression above, we decide to let x represent "five," then we can substitute 5 for x and have the expression $5 + 5 = 8$, which is not a true equation.

No, there is only one number that can be represented by x in the expression if we are to make an equation out of it, and that is "three." If we substitute 3 for x, we have the expression $3 + 5 = 8$, which is an equation. No other number substituted for x will do.

Of course, that is only what x is equal to in this particular expression. It may equal something else entirely in other expressions.

If $x + 17 = 19$, then $x = 2$; if $x + 8 = 13$, then $x = 5$, and so on. In each case you must pick the one number for x that makes an equation out of the expression.

SOLVING FOR x

But how do you pick a proper number for x when the equation becomes complicated? It is easy to see that x must be equal to 3 in the expression $x + 5 = 8$, because we know at once and with hardly any thought that $3 + 5 = 8$. But suppose we had the expression $x + 1865 = 2491$. How do we pick the proper value of x in that case?

We could try different numbers one after the other and wait until we happened to hit one that would make an equation out of the expression. If we were lucky, we might eventually happen to light on the number 626. If we substitute it for x, we have $626 + 1865 = 2491$, and behold, this is an equation. Hurrah! We now know that $x = 626$, and we have solved the equation.

Mathematicians, however, hate to use hit-and-miss tactics as a method of solving an equation. It's too uncertain and takes too long. Besides, there are methods for solving an equation that are *not* hit-and-miss. There are *rules* for solving equations.*

The rules tell you how to rearrange an equation so that it becomes easier to solve for x. There are numerous ways of rearranging an equation, but there is one thing you must always be careful of. In rearranging an equation, you must always keep it a true equation! Whatever you do, you must always see to it that the symbols on the left side of the equals sign represent the same quantity as those on the right side.

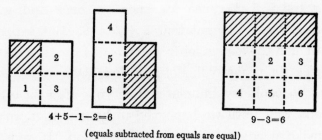

$$4+5-1-2=6$$

$$9-3=6$$

(equals subtracted from equals are equal)

---

* About 825, some of these rules were first presented in a book written by an Arabian mathematician named Mohammed ibn Musa al-Khowarizmi. The name of his book, in Arabic, is "ilm al-jabr wa'l muqabalah," which means, in English, "the science of reduction and cancellation." Reduction and cancellation were the methods he used to deal with equations, you see. Al-Khowarizmi didn't use the symbols we use today, but his methods for dealing with equations so impressed Europeans when they first obtained translations of his book that the subject of handling equations is still called "algebra," which is a mispronunciation of the second word in the book's title.

One way of rearranging an equation without making it false is to add the same quantity to both sides, or to subtract the same quantity from both sides. We can see examples of this clearly if we use only numerical symbols. For instance, since $3 + 5 = 8$, then

$$3 + 5 - 4 = 8 - 4$$

and

$$3 + 5 + 7 = 8 + 7$$

In the first case, both sides of the equation equal four; in the second, both sides equal fifteen.

Well, anything that applies to numerical symbols applies also to literal symbols. (This is the key to understanding algebra.) If we say that $x + 5 = 8$ is an equation, then $x + 5 + 3 = 8 + 3$ is also an equation, and so is $x + 5 - 2 = 8 - 2$.

Now suppose, in this particular case, we subtract five from each side of the equation. We begin with $x + 5 = 8$ and, subtracting five from each side, we have

$$x + 5 - 5 = 8 - 5$$

But if we add five to any quantity, then subtract five, we are left with the quantity itself. It's like taking five steps forward, then five steps backward; we end where we started. Thus, $3 + 5 - 5 = 3$; $17 + 5 - 5 = 17$, and so on.

Consequently, $x + 5 - 5 = x$, and when we say that $x + 5 - 5 = 8 - 5$, we are actually saying $x = 8 - 5$.

What we have worked out, then, is this:

If $\qquad x + 5 = 8$

then $\qquad x \qquad = 8 - 5$

We seem to have rearranged the equation by shifting the 5 from the left side to the right side. Such a shift is called a "transposition" (from Latin words meaning "to put across"), but please be very careful to notice how it came about. We didn't really move the 5; what we did do was to subtract a 5 from both sides of the equation, and the effect was as though we had moved the 5.

Nowadays, mathematicians like to concentrate on subtracting equal numbers from (or adding equal numbers to) both sides of an equation and let what seems to be the transposition take care of itself. However, as one gets used to handling equations, it begins to seem a waste of time always to add and subtract numbers when the same result arises by just shifting a number from one side of the equation to the other. I will do this throughout the book and I will talk about "transposing" and "transposition." I hope you will continue to think of such a way of treating equations as nothing more than a short cut, and remember that what I am really doing is subtracting equal numbers from (or adding equal numbers to) both sides of the equation.

Notice also that when I transposed the 5 in the equation I used above as an example, the plus sign changed to a minus sign. This shows one of the dangers of shifting numbers without stopping to think of what you are really doing. If you merely shift a 5, why should the plus sign be affected? But if you subtract 5 from both sides of the equation, then the plus sign automatically becomes a minus sign as the 5 seems to shift.

What if we had started with the equation

$$x - 5 = 8$$

We can add 5 to both sides of the equation and keep it true, so that $x - 5 + 5 = 8 + 5$. But since $x - 5 + 5 = x$ (if you go backward five steps, then forward five steps, you end in the starting place) then

$$x = 8 + 5$$

Again, it is as though we had shifted, or transposed, the 5, and again we have changed the sign, this time from minus to plus.

Now addition, represented by the plus sign, and subtraction, represented by the minus sign, are examples of "operations" performed upon numbers, whether represented by numerals or by letters.

These operations are constantly used in arithmetic and also in algebra, so they may be called "arithmetical operations" or "algebraic operations." We are concentrating on algebra in this book, so I will speak of them as algebraic operations.

Addition and subtraction, taken together, are examples of "inverse operations," meaning that one of them undoes the work of the other. If you add five to begin with, you can undo the effect by subtracting five afterward. Or if you subtract five to begin with, you can undo that by adding five afterward. We have just had examples of both.

You can see then that transposition changes an operation to its inverse. An addition becomes a subtraction on transposition, and a subtraction becomes an addition.

Do you see why all this should be helpful? Let's go back to the big-number equation I used near the beginning of the chapter. It was

$$x + 1865 = 2491$$

By transposition, the equation becomes

$$x = 2491 - 1865$$

The expression on the right side of the equation

is now ordinary arithmetic. It can be easily worked out to 626, so we can write

$$x = 626$$

and we have solved for x not hit-and-miss, but by the smooth working out of an algebraic rule. In the same way, the equation $x - 3489 = 72$ becomes, by transposition, $x = 72 + 3489$, so that x works out to be equal to 3561.

But how can you be sure you have solved the equation? How can you feel certain that you can trust the rules of algebra? Whenever you have obtained a numerical value for x in however complicated an equation and by however complicated a method, you should be able to substitute that numerical value for x in the original equation without making nonsense of it.

For the equation $x - 3489 = 72$, I have just worked out the value of x to be 3561. Substituting that for x in the equation, we have $3561 - 3489 = 72$. Common arithmetic shows us that this is an equation, so our value for x is correct. No other numerical value would have made an equation out of this expression.

Now that we have a rule, how is it best to state it? I have been using particular numbers. I have said that if you begin with $x + 5 = 8$, you can

change that to $x = 8 - 5$, and if you begin with
$x + 1865 = 2491$, you can change that to $x = 2491 - 1865$. However, when you put a rule that
way there is always the danger that you might be
giving the impression that the rule holds only for
the particular set of numbers you have used as
an example.

One way of trying to avoid that would be to list
the rule for all possible sets of numbers, but no one
would be foolish enough to try that. There is an
endless group of sets of numbers and such a task
would never be finished. Instead, we can turn to the
use of symbols again.

Suppose we let a and b stand for a pair of numbers.
They might stand for 1 and 2, or for 3 and 5, or for
75 and 8,358,111 — any pair of numbers at all.
Then we can say that one rule covering the handling
of equations is this:

$$\text{If} \qquad x + a = b$$
$$\text{then} \qquad x = b - a$$
$$\text{And if} \qquad x - a = b$$
$$\text{then} \qquad x = b + a$$

Now, you see, the rule covers not just particular
sets of numbers, but any set. We have used general
symbols, instead of particular numerals.

BOTH SIDES OF ZERO

Have I gone too far now? I have said that a and b can stand for any numbers, so suppose I let a stand for 8 and b stand for 3. Now the general equation $x + a = b$ becomes $x + 8 = 3$. The rule of transposition lets the equation be changed to $x = 3 - 8$, and the question is: What does $3 - 8$ mean?

The early mathematicians considered expressions of the type of $3 - 8$ to have no meaning. How can you take eight away from three? If you only have three apples, how can anyone take eight apples from you? What they decided, then, was that an equation such as $x + 8 = 3$ had no solution for x and they refused to work with such equations.

This will not do, however. Mathematicians hate to be in a position where they are faced with an unknown for which they can find no solution. Sooner or later, one of them will work up a system which will allow a solution. In this case, the mathematician to do so was an Italian named Geronimo Cardano (kahr-DAH-no), back about 1550.

The system is simple enough. If you have three apples, it *is* possible for someone to take eight apples from you. All that has to happen is for you to give him the three you have and agree to owe him five more. You must assume a debt.

In the same way, if you are at the starting post of a race, take three steps forward, then eight steps backward, you end up five steps behind the starting post.

All you need is some way of showing numbers that are less than zero; that represent a debt; that mark a position behind the starting post. Cardano pointed this out carefully. Since such numbers are come across in the process of subtraction, the minus sign is used to distinguish them from ordinary numbers.

Thus $3 - 8 = -5$. If $x = 3 - 8$, then $x = -5$. Cardano's system produced a reasonable solution for x in such cases.

Numbers with a minus sign, which symbolize quantities less than zero, are called "negative numbers." The word "negative" comes from a Latin word meaning "to deny." That shows how reluctantly mathematicians came to use such numbers even after they realized they had to, unless they wanted to leave certain equations unsolved. It was as though they were still denying that such numbers really existed.

Ordinary numbers, symbolizing quantities greater than zero, are called "positive numbers." When you want to show beyond the shadow of a doubt that a number is a positive number and not a negative one, you can put a plus sign before it. This is done because, in the process of addition, only positive numbers ever arise out of positive numbers. Instead of writing simply 5, you might write +5.

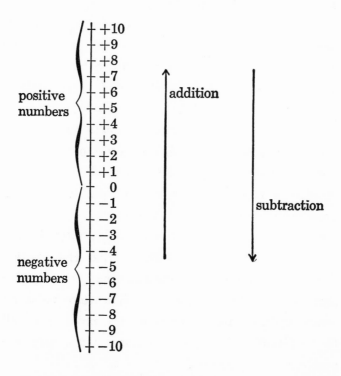

Positive numbers, however, were used for so many centuries before negative numbers were accepted — and, what's more, positive numbers are still used so much more than are negative numbers — that everybody takes the plus sign for granted. Whenever you see a number without a sign, you can safely assume it is a positive number. This means that if you want to use a negative number, you *must* put a minus sign before it, or everyone will take it for a positive number.

There is one drawback to this particular system of signs and that is that we are making the same sign do two different jobs, which can be confusing. The sign $+$ is used to indicate the operation of addition, in which case it is properly called the "plus sign." It is also used to indicate a number to be positive, in which case it should be called the "positive sign." In the same way, the minus sign should be called the "negative sign" when it is used to indicate a negative number.

One reason why we can get away with letting these symbols do double duty is that the positive sign (which is used so much more than the negative sign) is generally omitted, so that we're not even aware of it. It makes us think the plus sign is all there is. But let's try to write an equation with the positive signs included.

For instance, the equation $x + 3 = 5$ should really be written

$$(+x) + (+3) = (+5)$$

We use the parenthesis to show that the + symbol inside it belongs to the number and is a positive sign. The parenthesis hugs symbol and number together, so to speak. The + symbol between the parentheses is a true plus sign and signifies the process of addition.

Let's see how this works out if we use negative numbers. Suppose we have the equation

$$(+x) + (-3) = (+5)$$

Well, when we add a negative number to something, we are adding a debt, so to speak. If I give you a three-dollar debt, I am adding that debt to your possessions, but that is the same as making you poorer by three dollars. I am taking three dollars from you. There the equation can also be written

$$(+x) - (+3) = (+5)$$

Very much the same thing happens, if we subtract a negative number, as in

$$(+x) - (-3) = (+5)$$

If I take a three-dollar debt away from you by offering to pay it myself, you are automatically three dollars richer. It is as though I had given you three dollars in cash. So this equation can be written

$$(+x) + (+3) = (+5)$$

Now let's make this as simple and as familiar in appearance as possible by leaving out the positive signs and the parentheses that go with them. Instead of saying that $(+x) + (-3)$ is the same as $(+x) - (+3)$, we will say that

$$x + (-3) = x - 3$$

This looks like an equation, but it is more than that. An ordinary equation, such as $x + 2 = 5$, is true only for a particular value of x; in this case only for $x = 3$.

The statement that $x + (-3)$ is the same as $x - 3$, however, holds true for all values of x. No matter how much money you have, in other words, adding a three-dollar debt is the same as taking away three dollars in cash.

Statements which hold true for all possible values of x are called "identities." Often, but not always, a special sign is used to indicate an identity, and I will make use of it whenever I want to show an identity. The "identity sign" consists of three short dashes, $\equiv$, a kind of reinforced equality, so to speak. It is read "is the same as" or "is identical with."

So we can write: $x + (-3) \equiv x - 3$, or $x - (-3) \equiv x + 3$.

To make the rule general, we should avoid using

particular numbers. It works for all numbers and we should therefore use general symbols, such as the letter a, and say

$$x + (-a) \equiv x - a$$

$$x - (-a) \equiv x + a$$

AVOIDING THE NEGATIVE

This way of switching from negative to positive comes in handy when negative numbers occur in equations and must be dealt with. Suppose you had the equation $x + (-3) = 5$. You would transpose the -3, changing the addition to a subtraction (it is the sign of the operation that is changed, not the sign of the number) and get $x = 5 - (-3)$.

This can at once be changed to $x = 5 + 3$, which comes out to 8.

Or you could tackle the equation before transposing. Keeping the rules of signs in mind, you could change $x + (-3)$ to $x - (+3)$, which, of course, you would write as simply $x - 3$, so that the equation becomes $x - 3 = 5$. Transposing, you would get $x = 5 + 3$, which again comes to 8.

You see, if you use algebraic rules properly, the value of x will always come out the same, no matter what route you take to arrive at that solution.

If you work out an equation one way and come out with $x = 6$, then work it out another way and come out with $x = 8$, you can be sure that you have made a mistake in handling the rules.

Sometimes, in working with complicated equations, it is not easy to see where a mistake in handling the rules was made. Then it seems as though one can use the rules to show something that is nonsense; that $1 = 2$, for instance. Such an apparent contradiction is called a "fallacy," from a Latin word meaning "deceive." Professional mathematicians working out new advances are particularly anxious to avoid fallacies, but once in a while even the best of them may fall victim to one.

The rules governing positive and negative signs offer another advantage. They make it possible to avoid subtraction altogether by changing all subtractions to additions. Instead of ever writing $x - 3$, we can always write $x + (-3)$.

The point in doing this is that some of the rules governing addition are not the same as those governing subtraction. I can show this by first considering operations involving numerical symbols only.

It doesn't matter, does it, in what order given numbers are added? Thus, $5 + 3 = 8$, and $3 + 5 = 8$. If Joe gives you $5 and Jim gives you

$3, you don't care which one pays up first; you end with $8 total either way. Using general symbols, you can say

$$a + b \equiv b + a$$

But how about subtraction? If $5 - 3 = 2$, does $3 - 5$ give you the same quantity? It does not. As you now know, $3 - 5 = -2$. The two answers are not the same and therefore

$$a - b \neq b - a$$

where, as you can probably guess, the symbol $\neq$ means "is not identical with."*

You see, then, that if you're handling only additions, you can relax as far as the order in which the symbols are taken is concerned. If you're handling subtractions, you have to be careful about the order, or you may find yourself coming out with the wrong answer.

This is not likely to happen in ordinary arithmetic, where you would probably never write $3 - 5$ when you mean $5 - 3$. However, in complicated equations where arrangements and rearrangements are constantly being made, it is only too

---

* Of course, if a and b represented the same number, then $a - b \equiv b - a$, because both expressions would equal 0. This, however, is not an important exception. It is what mathematicians call "trivial."

easy to shift the order of the various symbols without noticing. You might then find yourself in the middle of a fallacy.

To avoid trouble, you might eliminate all subtractions by changing expressions like $a - b$ to $a + (-b)$. Then the order makes no difference. If you write an expression as $x - 3$, for instance, you must never write it as $3 - x$. If you write it instead as $x + (-3)$, you can write it as $(-3) + x$ with complete ease of mind.

This method of changing subtraction to the addition of negative numbers is called "algebraic addition" simply because one first comes across it in algebra. It is really no different from ordinary addition, however, once you have learned to handle positive and negative signs.

Incidentally, when a negative number is the first symbol in an expression, it is not necessary to use the parenthesis since there is no sign for any operation preceding it, and no chance of confusing the negative sign with a minus sign. For that reason, $x + (-a)$ is always written with the parenthesis, but $(-a) + x$ is usually written as simply $-a + x$.

3

More Old Friends

IN THE previous chapter, all I talked about were two algebraic operations, addition and subtraction. Now it is time to pass on to two more, multiplication and division. These two operations are also old friends, frequently used in ordinary arithmetic, and used in exactly the same way in algebra.

To show you how these new operations might arise as part of equations, suppose that you buy a number of oranges for 48¢. You don't know exactly how many oranges there are in the bag, but you know that these particular oranges are 4¢ apiece. Therefore, you know that the number of oranges, whatever that number is, multiplied by 4, will give the answer 48. If you let the number of oranges be represented by x, you can write the equation

$$x \times 4 = 48$$

Now we need to solve for x. But how does one

go about this where the multiplication sign ($\times$) is involved?

In the previous chapter, remember, I said that an equation remains an equation if the same number is added to both sides or subtracted from both sides. Well, it is also true that an equation remains a true equation if both sides are multiplied by the same number. For that matter, both sides may also be divided by the same number (with the one exception that neither side can be divided by zero, as I shall explain shortly). Suppose, for instance, we divide both sides of the equation just given by 4. We can write the result this way, then:

$$x \times 4 \div 4 = 48 \div 4$$

where $\div$ is the symbol for division.

Now if we multiply a number by a particular quantity, then divide the product by that same quantity, we are back to the original number. (If you're in doubt, try it on various numbers.) Or, to put it another way, the expression $4 \div 4$ is equal to 1, so that $x \times 4 \div 4$ is equal to $x \times 1$. Then, since any number, known or unknown, multiplied by 1, remains unchanged, $x \times 1$ is equal to x.

The equation becomes, therefore,

$$x = 48 \div 4 \qquad \text{(or 12)}$$

A very similar situation works itself out if we have an equation involving a division such as

$$x \div 4 = 12$$

Suppose we multiply both sides of the equation by 4. Then we have

$$x \div 4 \times 4 = 12 \times 4$$

Dividing by 4, then multiplying the quotient by 4, gives us back the original x, of course, so that the equation becomes

$$x = 12 \times 4 \qquad \text{(or 48)}$$

Since division undoes the effect of multiplication, and vice versa, these two are inverse operations and form a pair after the fashion of addition and subtraction.

If you go over what has been done so far in this chapter, you will see that, just as in the case of addition and subtraction, handling multiplications and divisions involves what seems a shift in numbers from one side of the equation to the other. This is so like the transposition I mentioned in the previous chapter that I will call this shift by the same name. This new kind of transposition is still a short cut and nothing more. Remember that what is really involved is multiplication (or division)

by equal numbers on both sides of the equation.
And again, this new transposition changes an operation into its inverse; multiplication is changed to
division, and division is changed to multiplication.
We can make a general rule of this by using
letter symbols.

$$\text{If} \qquad x \times a = b$$
$$\text{then} \qquad x \quad = b \div a$$
$$\text{And if} \qquad x \div a = b$$
$$\text{then} \qquad x \quad = b \times a$$

In dealing with division, by the way, mathematicians have a special rule that is quite important,
and I had better tell you about it now. This rule
absolutely forbids division by zero. Dividing by
zero makes no sense, you see, for ask yourself what
$5 \div 0$ is, or $10 \div 0$, or any number, for that matter,
divided by zero. That's like asking how many
zeros must be lumped together in order to reach 5,
or 10, or any number. It's like asking how many
times you must put nothing into a pail in order to
fill it. These are senseless questions, you see, and
to avoid trouble, they should not even be asked.

In arithmetic, it is easy not to divide by 0, but in
algebra, there are dangers. Sometimes a number is
divided by a combination of literal symbols that
happens to equal zero without the mathematician

noticing. If ordinary rules of algebra are applied to such an expression, nonsensical answers are often produced. In fact, a very common source of fallacies is the accidental division by zero somewhere in the manipulation of the equations.

Consequently, when I use an expression such as $x \div a = b$, b can represent any value at all, as is usual with literal symbols. The symbol a, however, can only represent any value *except zero*. This must be kept in mind.

ELIMINATING THE CONFUSION IN SYMBOLS

Actually, the multiplication and division signs, although common in arithmetic, are hardly ever used in algebra. For one thing, the multiplication sign is very much like an x. This symbol of the unknown is not used in ordinary arithmetic, so there is no danger of confusion there. In algebra, however, where x is used in almost every equation, the possibilities of confusion between x and $\times$ are very good.

There are other ways in which the operation of multiplication is often symbolized. One is by the use of a dot. Instead of writing $x \times 4$, we could write $x \cdot 4$. This is often done in algebra, but hardly ever in ordinary arithmetic, for to write $2 \cdot 3$ instead of 2×3 is to raise the possibility of confusion with the decimal point. To be sure, the decimal point is written at the bottom of the line, 2.3, while the multiplication dot is centered, 2·3, but if you are reading rapidly you are very likely not to notice the difference.

A still greater simplification is that of doing away with a symbol for multiplication altogether. Suppose you say "four apples." What you really mean is "four times one apple." Similarly, you can speak of 4 pairs or 4 dozen or 4 anything.

You could even write or speak of 4 6's. By that you would mean 6 and 6 and 6 and 6, and if you are interested in the total quantity, that is 4 × 6. To speak of 4 6's is therefore much the same as speaking of 4 × 6. The difficulty of doing this in ordinary arithmetic, however, is that to write 4 6's is to risk a great chance of confusion with the number 46.

In algebra, however, you can easily write 4 x's to indicate 4 × x. There's no danger of confusion there. You could even bring the 4 and the x right up next to each other without leaving any space and still have no confusion; and you can leave out the plural (saying "four eks" rather than "four ekses"). In other words, 4 × x can be written simply 4x and, in algebra, this is almost always done. It will be done from here on, in this book.

In order to use this very convenient system even when only numerals are involved, parentheses can be used. Each number involved in the multiplication is enclosed in a parenthesis showing 'that it is a number all by itself; thus, (4)(6) or (46)(23). The numbers are thus kept apart and cannot be 46 in the first place or 4623 in the second. Sometimes this is made more emphatic by using the multiplication dot and writing (4)·(6), but this is not really necessary.

I will indicate the multiplication of numerical symbols by means of parentheses in this book from now on, in order to avoid the multiplication sign. This may seem strange to you at first, but you will quickly grow used to it.

The ordinary division sign in arithmetic also allows room for confusion. It is too like the minus sign, differing by only two dots which are easily overlooked. And if the two dots smudge a little, the sign can become similar to the plus sign.

In algebra, then, it is usual to indicate division by drawing the two symbols together, with a line between. The line may be either slanting or horizontal. The slanting line, /, is sometimes called a "shilling mark" because the British use it as a symbol for their coin, the shilling. The horizontal line, —, used in division may look even more like a minus sign than does the ordinary division sign, but there are crucial differences. A minus sign lies between two numbers, one on its left and one on its right. The horizontal division line separates two numbers above and below, and there is actually no danger of confusion at all.

Thus $x \div 12$ can be written $x/12$ or $\frac{x}{12}$. These new symbols can also be used where only numerals are involved. Instead of $48 \div 4$, you can write $48/4$ or $\frac{48}{4}$.

Using this system for symbolizing multiplication and division, we can write the general rules for handling them in equations as follows:

$$\text{If} \quad ax = b$$

$$\text{then} \quad x = \frac{b}{a}$$

$$\text{And if} \quad \frac{x}{a} = b$$

$$\text{then} \quad x = ba$$

Compare these with the rules given on page 34 and you will see that I have only changed the system of indicating the operations and nothing more.

Let me now explain something about multiplication and division that resembles a point I have already made in connection with addition and subtraction.

When you multiply two numbers together, it

doesn't matter which number you multiply by which. Thus, $(6)(8) = 48$ and $(8)(6) = 48$ also. This is a general rule that can be written

$$ab \equiv ba$$

It follows then that $4x \equiv x4$, yet although these two expressions have identical values, mathematicians always write $4x$ and never write $x4$. It isn't incorrect to write $x4$; it just isn't done. You might almost think of it as a kind of mathematical etiquette, like not using the wrong fork for salad, even though you can eat the salad easily with it.

Whenever mathematicians, or any other group of people, in fact, all make use of a particular way of doing things when another way might do just as well, they are adopting a "convention." For instance, any letter, such as q or m or even a made-up sign such as $\boxed{+}$ would do to represent an unknown quantity, but it is conventional, the world over, to use x.

Such conventions are by no means a sign of sheeplike behavior. They are an important convenience. They make certain that all mathematicians everywhere speak the same mathematical language. It would be troublesome, time-wasting, and a source of confusion to have one mathematician puzzled by the writings of another just because each was using a different convention. They would both be right, perhaps, but neither would be clear to the other.

As for division, here, as in the case of subtraction, the order of the number does make a difference. The expression $\frac{15}{3}$ has the value of 5, while the expression $\frac{3}{15}$ has the value $\frac{1}{5}$. We can express this generally by saying

$$\frac{a}{b} \neq \frac{b}{a}$$

MANEUVERING FRACTIONS

We seem to have stumbled into fractions* here, for certainly expressions such as $\frac{x}{12}$ or $\frac{x}{a}$ look like fractions. And, as a matter of fact, they are fractions.

The solution of any equation involving a multiplication or division is quite likely to introduce a fraction. Sometimes such a fraction can be con-

---

* I am going to assume in this book that you know how to handle fractions and decimals in ordinary arithmetic. If, by any chance, you feel a little shaky or just want to brush up on general principles, you could glance through Chapters 4 and 5 of *Realm of Numbers*.

verted to an ordinary "whole number" or "integer." Thus, earlier in the chapter, we came up against the fraction $\frac{48}{4}$, which can be written as the whole number 12. This is the exceptional case, however. Suppose, instead, that we have the equation

$$4x = 47$$

By transposing, we have

$$x = \frac{47}{4}$$

The fraction, $\frac{47}{4}$, cannot be changed into a whole number. The best you can do is write it as $11\frac{3}{4}$.

A fraction can also be referred to as a "ratio," and it is important to remember that whole numbers can also be written as fractions, or ratios. The number 12 can be written as $\frac{12}{1}$, or $\frac{24}{2}$, or $\frac{48}{4}$. For that reason, whole numbers and fractions, both positive and negative, are lumped together as "rational numbers," that is, numbers which can be expressed as fractions, or ratios.

Fractions in algebra are handled in the same way as in arithmetic. For instance, equations involving the addition and subtraction of fractions introduce nothing new at all.

If $\qquad x + 2\frac{1}{2} = 6$

then $\qquad x = 6 - 2\frac{1}{2} \qquad \left(\text{or } 3\frac{1}{2}\right)$

And if $\qquad x - \frac{1}{4} = 3$

then $\qquad x = 3 + \frac{1}{4} \qquad \left(\text{or } 3\frac{1}{4}\right)$

(It might suddenly occur to you that by the algebraic system of denoting multiplication, a number like $3\frac{1}{4}$ might signify 3 multiplied by $\frac{1}{4}$. However, it doesn't. In arithmetic, $3\frac{1}{4}$ means 3 plus $\frac{1}{4}$, and algebra accepts that as too familiar to change. To write 3 multiplied by $\frac{1}{4}$ without using the multiplication sign, parentheses must be used thus, $(3)\left(\frac{1}{4}\right)$. Where literal symbols are used, we need not be so careful, since there is nothing in ordinary arithmetic for $x\frac{1}{4}$ to be confused with. That expression means x multiplied by $\frac{1}{4}$, but, of course, it is always written $\frac{1}{4}x$ by convention. If you do want to indicate an addition, you write it out in full,

$$x + \frac{1}{4}.\Big)$$

Now suppose you find yourself involved in the multiplication of a fraction. Such a situation might arise as follows. You are told that your share of a certain sale will amount to $\frac{2}{5}$ of the total. The sale is made and you are given \$18. From that you can calculate what the total sale amounted to. If you let the unknown value of the total sale be represented by x, then $\frac{2}{5}$ of that is 18 and you can write the equation

$$\frac{2}{5}x = 18$$

There are several ways of proceeding. First, any fraction multiplied by its reciprocal* is equal to 1.

---

* The reciprocal of a fraction is another fraction in which the numerator and denominator have exchanged places. For example, the reciprocal of $\frac{2}{3}$ is $\frac{3}{2}$; the reciprocal of $\frac{5}{14}$ is $\frac{14}{5}$, and so on. Even whole numbers have reciprocals, for since 12 can be written $\frac{12}{1}$, its reciprocal is $\frac{1}{12}$. And a reciprocal can be written as a whole number, too, for the reciprocal of $\frac{1}{7}$, for instance, is $\frac{7}{1}$, which can be written simply as 7.

For instance, $\left(\dfrac{2}{5}\right)\left(\dfrac{5}{2}\right)$ is equal to $\dfrac{10}{10}$, which, in turn, equals 1.

Suppose, then, that we multiply both sides of the equation by $\dfrac{5}{2}$, thus:

$$\left(\frac{2}{5}\right)\left(\frac{5}{2}\right) x = (18) \left(\frac{5}{2}\right)$$

Since $\left(\dfrac{2}{5}\right)\left(\dfrac{5}{2}\right)$ equals 1, the left-hand side of the equation becomes simply x, and by working out the right-hand side in ordinary arithmetic, we find that x equals 45. That is the full value of the sale, for if you check you'll find that $\dfrac{2}{5}$ of 45 is indeed 18.

The general rule for such a situation, then, is that a fraction involved in a multiplication, when transposed, is converted to its reciprocal, thus:

If $\qquad\left(\dfrac{a}{b}\right) x = c$

then $\qquad x = c\left(\dfrac{b}{a}\right)$

It may seem to you that here is a case where a multiplication of a fraction is left a multiplication of a fraction after transposition, instead of being converted to the inverse operation. However, there is an explanation for this. An expression like $\left(\dfrac{a}{b}\right)x$

involves both a multiplication and a division since it stands for a multiplied by x and then the product divided by b. (If you doubt this, check an expression involving numerals, such as $\left(\dfrac{2}{3}\right)(6)$ and see if the answer isn't obtained by working out 2 multiplied by 6 and then the product divided by 3.) When the fraction is transposed, the multiplication becomes a division and the division becomes a multiplication. It is because of this *double* change to the inverse that there seems to be no change at all.

This can be made plainer by handling the fraction one piece at a time. Since $\left(\dfrac{2}{5}\right)x$ signifies 2 multiplied by x divided by 5, it can be written $\dfrac{2x}{5}$, and the equation $\left(\dfrac{2}{5}\right)x = 18$ becomes

$$\frac{2x}{5} = 18$$

You can now transpose the 5 in accordance with the rules and have

$$2x = (18)(5) \qquad \text{(or 90)}$$

and then transpose the 2, so that

$$x = \frac{90}{2} \quad \text{(or 45)}$$

Or, if you preferred, you could first transpose the 2 to give

$$\frac{x}{5} = \frac{18}{2} \quad \text{(or 9)}$$

and then the 5, so that

$$x = (9)(5) \quad \text{(or 45)}$$

Or perhaps you would like to work with decimals. The fraction $\frac{2}{5}$ is equal to 0.4 in decimals, so you can say

$$0.4x = 18$$

If you multiply both sides of the equation by 10, you can get rid of the decimal point since $(0.4)(10)$ is equal to 4. Therefore you have

$$4x = 180$$

And, by transposing

$$x = \frac{180}{4} \quad \text{(or 45)}$$

The important point here, once again, is that no matter what rules you use for handling the equation, *as long as the rules are correct and are correctly used*, you will always come out with the same value for x. In the example I gave you here, x comes out equal to 45, no matter how the symbols were manipulated.

WATCHING THE SIGNS

How do negative numbers fit into equations involving multiplication and division? Suppose we have

$$\frac{x}{3} = -5$$

By transposing, we find

$$x = (-5)(3)$$

and what does that mean as far as the value of x is concerned?

Now $(-5)(3)$ indicates a multiplication of -5 by 3. It is the equivalent of tripling a five-dollar debt. If three people each have a five-dollar debt, the total for the group is a fifteen-dollar debt. Therefore $(-5)(3)$ is equal to -15 and that is the value of x in the equation above. If you had taken five debts of three dollars each, you would still have ended with a fifteen-dollar debt, so $(5)(-3)$ is also equal to -15.

Using letter symbols to make the rule general:

$$(+a)(+b) \equiv +ab$$
$$(-a)(+b) \equiv -ab$$
$$(+a)(-b) \equiv -ab$$

This does not use up all the variations that are possible. What if both numbers being multiplied are negative? For instance, in the equation $\dfrac{x}{(-3)} =$ -5, transposition shows that $x = (-5)(-3)$. How do you evaluate x now?

Unfortunately, there is no easy way of seeing the meaning of such a multiplication of a negative by a negative. It might represent a debt of five dollars held by each of -3 people, but what on earth can we mean by -3 people?

Instead of trying that, let's take a closer look at the three rules for multiplication of signs I have already given you. Notice that when a quantity is multiplied by a positive number, the sign of the product is the same as the sign of the original quantity. If $+a$ is multiplied by a positive number, the sign of the product is $+$; while if $-a$ is multiplied by a positive number, the sign of the product is $-$.

It sounds reasonable to suppose that when a quantity is multiplied by a negative number, the sign of the product is the reverse of the sign of the original quantity. We have one case in the three rules where $+a$ is multiplied by a negative number and the sign of the product is $-$. If that is so, then when $-a$ is multiplied by a negative number, the sign of the product should be $+$. This conclusion has proven satisfactory to mathematicians, so we can say

$$(-a)(-b) \equiv +ab$$

The same rule of signs holds in division as in multiplication. This can be shown in several different ways, but I shall do so by making use of reciprocals.

From ordinary arithmetic, we know that $\frac{10}{2} = 5$ and that $(10)\left(\frac{1}{2}\right) = 5$. Notice, too, that $\frac{1}{2}$ is the reciprocal of 2 (which can be written, remember, as $\frac{2}{1}$). Again, $\frac{15}{5}$ and $(15)\left(\frac{1}{5}\right)$ both come out to be equal to 3, and $\frac{1}{5}$ is the reciprocal of 5. For that matter, $\frac{10}{\left(\frac{2}{5}\right)}$ will give you the same answer as $(10)\left(\frac{5}{2}\right)$. It is 25 in both cases. Perhaps you can see this more easily if you change the fractions into decimals. The fraction, $\frac{2}{5}$ is 0.4, while $\frac{5}{2}$ is 2.5. Well, $\frac{10}{0.4}$ and $(10)(2.5)$ both come out to 25.

In short, you can try any number of such cases and you will always find that it doesn't matter whether you divide a quantity by a particular number or multiply that quantity by the reciprocal of that particular number. Either way, you get the same answer. Speaking generally:

$$\left(\frac{a}{b}\right) x \equiv \frac{x}{\left(\dfrac{b}{a}\right)}$$

This means that just as we can always turn a subtraction into an addition by changing the sign of the number being subtracted, so we can always turn a division into a multiplication by taking the reciprocal of the divisor.

If we have the expression $\dfrac{(-15)}{5}$, which involves the division of a negative number by a positive number, we can change it to $(-15)\left(\dfrac{1}{5}\right)$, which involves the multiplication of a negative number by a positive number. Since $(-15)\left(\dfrac{1}{5}\right)$ must equal -3 by the rule of signs (negative times a positive equals a negative), then $\dfrac{(-15)}{5}$ must also equal -3.

Thus, the rule of signs must be the same in division as in multiplication. If it were not, we would be stuck with two different answers according to the method we used to obtain those answers. We might get a $+3$ if we divided, but a -3 if we used the reciprocal rule and then multiplied.

The rule of signs would then be "inconsistent" with the reciprocal rule, and this is a fatal sin in mathematics. Mathematicians feel they must be "consistent" at all costs. All their rules must fit together, and no one rule must contradict any other.

In the interest of consistency, then, the rule of signs in divisions can be expressed thus:

$$\frac{+a}{+b} \equiv +\frac{a}{b}$$

$$\frac{-a}{+b} \equiv -\frac{a}{b}$$

$$\frac{+a}{-b} \equiv -\frac{a}{b}$$

$$\frac{-a}{-b} \equiv +\frac{a}{b}$$

4

Mixing the Operations

MORE THAN ONE

So FAR, I have kept my equations as simple as I can. I have used no expression with more than one plus sign or one minus sign in it. There is no rule, though, that makes this necessary. I have complete liberty to write an expression such as $x + 3 + 2 - 72$, or one of any length, if I wished.

Each of the items being added or subtracted in such an expression is called a "term." It doesn't matter whether the item is a numerical symbol or a literal symbol. In the expression I have just used, 72, 2, 3, and x are all terms.

An expression which is made up of a single term is called a "monomial." (The prefix "mono-" is from the Greek word for "one.") Expressions with more than one term are named by the use of prefixes representing the particular number (in Greek) of terms involved. An expression with two terms, such as $x + 3$, is a "binomial," one such as $x + 3 + 2$ is a "trinomial," one such as $x + 3 + 2 - 72$ is a "tetranomial," and so on. It is usual, however, to lump together all expressions containing more than one term as "polynomials," where the prefix "poly-" comes from the Greek word for "many."

In ordinary arithmetic, little attention is paid to the number of terms in any expression, since by adding and subtracting they can all be reduced to a single term anyway. Faced with an expression like $17 + 5 - 16 + 12 - 3$, it is the work of a moment (thanks to a few years of painful drill in the first few grades) to convert it into the single term 15.

In algebra, however, where literal terms are involved, matters aren't quite that simple. All is not lost, though. For one thing, all the numerical terms, at least, can be combined into a single term. The equation $x + 3 + 2 + 5 = 17 + 4 - 9$ can be changed without trouble to $x + 10 = 12$.

As for literal terms themselves, where more than one is involved in a particular expression, something can be done where only one kind of literal symbol is involved. If you are faced with $x + a$, to be sure, there is no way of performing the addition until you have decided what quantities x and a stand for. If, however, you are faced with $x + x$, you don't have to know what x stands for.

For any value of $x, x + x$ is $2x$. It should be stated as an identity: $x + x \equiv 2x$.

In the same way, $3x + 2x \equiv 5x$, and $75x - 11x \equiv$

64x. For that reason, an equation like this

$$5x - 3x + 2x = 11 + 12 - 7$$

works out at once to

$$4x = 16$$

The terms 2x and 64x, and others like them, consist of two parts, one numerical and one literal. It is customary for mathematicians to refer to the numerical part as the "coefficient," a word first used for this purpose by no less a person than Vieta, the father of algebra.

Letter symbols can also be considered coefficients, so that in the term ax, a is the coefficient. The coefficient is always considered as being involved with the unknown by way of a multiplication, never by way of a division. A term that involves a division must be converted to a multiplication by the rule of reciprocals before you are safe in deciding on the coefficient. For instance $\frac{x}{2}$ should be written $\left(\frac{1}{2}\right) x$ and then you will see that $\frac{1}{2}$ is the coefficient and not 2.

As a matter of fact, even the expression x can be considered as having a coefficient. After all, it can be written 1x just as a book can be referred to as 1 book. The coefficient of x is therefore 1.

Where the coefficient is 1, it is generally omitted so that you never write $2x - 1x = 1x$, but always $2x - x = x$. But the coefficient is there just the same and it shouldn't be forgotten, because we'll have occasion to think of it before the book is done.

But now a thought may occur to you. A term like $2x$ involves an operation, that of multiplication. An expression such as $2x + 3$ involves two operations, one of multiplication and one of addition. It is 2 times x plus 3.

Does that mean that three terms are involved? When I first spoke of terms, I mentioned them as items that were being added or subtracted. What about items that are being multiplied or divided?

To answer these questions will require us to look into these algebraic operations a little further.

WHICH COMES FIRST?

We have already decided that when two quantities are added, it makes no difference which is added to which; in other words, that $a + b \equiv b + a$. But what if more than two quantities are being added? If you try, you will see that in an expression like $8 + 5 + 3$, your answer will be 16, in whatever order you take the numbers.

In fact, you take it for granted that order makes no difference in such cases. When you add a long column of figures from the top down, you can check the results by adding it a second time from the bottom up. You fully expect to get the same result either way, provided you make no arithmetical mistake, even though you've reversed the order the second time.

If that is so, then the order doesn't matter in the case of subtractions either, or where additions and subtractions are combined, *provided* all the subtractions are converted to additions by the use of negative numbers, according to the method I described in Chapter 2. Thus, although $6 - 3 \neq 3 - 6$, it is nevertheless true that $6 + (-3) = (-3) + 6$.

You can go through the same thing with multiplications and divisions. Try working out expressions such as $(8)(4)(2)$ and compare the answer you get with that obtained in expressions such as $(4)(8)(2)$, $(4)(2)(8)$, $(8)(2)(4)$, $(2)(4)(8)$, and $(2)(8)(4)$. It will be 64 in every case. If divisions, or multiplications combined with divisions are considered, the same thing holds. The order doesn't matter, provided the divisions are converted to multiplications by the use of reciprocals, according to the method I described in Chapter 3. Thus, although $\dfrac{x}{3} \neq \dfrac{3}{x}$, yet $x\left(\dfrac{1}{3}\right) \equiv \left(\dfrac{1}{3}\right)x$.

The general rule is that the order in which operations are performed does not matter where only additions are involved, or where only multiplications are involved.

The next question, though, is, What happens if an expression contains operations that are not only additions and are not only multiplications, but contain some of both?

Let's take the simplest case, an expression that involves one multiplication and one addition, and let's use only numerical symbols to begin with. In order to make things as clear as possible, I will temporarily return to the use of the multiplication sign.

The expression we can consider is $5 \times 2 + 3$. If we work out the operations from left to right, we find that 5×2 is 10, and that $10 + 3$ is 13. We might want to check that answer by the same method we use in checking the addition of columns of figures; that is, by working it backward to see if we get the same answer. Well, if we work it backward, $3 + 2$ is 5, and 5×5 is 25.

The answers are not the same, since the expression works out to 13 in one direction and 25 in the other. There are no arithmetical errors here, but we can't allow answers to change just by varying the working methods of solving a problem. That would be inconsistency. It is important to set up some sort of system that will prevent that from happening.

The system now used (first decided upon about 1600) is to enclose in parentheses those operations that ought to be performed first. The expression $5 \times 2 + 3$ might be written, for instance, $(5 \times 2) + 3$, in which case the multiplication is carried out first so that the expression becomes $10 + 3$ (with the parentheses disappearing once the operation has been performed) or 13. If, instead, the expression is written $5 \times (2 + 3)$, the addition is performed first and the expression becomes 5×5, or 25.

Now there is no inconsistency. Instead of a single expression, we have two expressions which, thanks to parentheses, can be written differently. Each expression has only one answer possible.

Very complicated expressions, including many additions and multiplications, can be handled by setting up parentheses within parentheses. Usually each set of parentheses is of a different shape for the sake of clarity, but all have the same function. The convention, then, is to perform the operations of the innermost parentheses first and proceed outward in order.

If you have the expression

$$\left\{ [4 + (6 \times 5) - 12] \times 4 \times \frac{1}{11} \right\}$$

the innermost parenthesis contains the expression

6 × 5, so that operation is performed first. It works out to 30 and, with that parenthesis gone, what remains is

$$\left\{ [4 + 30 - 12] \times 4 \times \frac{1}{11} \right\}$$

The innermost parenthesis of those remaining now contains the expression $4 + 30 - 12$. Only addition and subtraction are involved and it is no great feat to reduce it to 22, so that what is left is

$$\left\{ 22 \times 4 \times \frac{1}{11} \right\}$$

Now only multiplications are involved, and the final value of the entire original expression is therefore 8.

Where only numerical symbols are involved, parentheses are easily removed by performing the operations within them. In algebra, with its literal symbols, parentheses cannot be removed that easily. That is why the matter of parentheses is more important in algebra than in arithmetic, and why you generally don't encounter parentheses in any important way until you begin the study of algebra.

Yet the same system applies to literal symbols as to numerical ones. No new complications are introduced. Consider the expression $6 \times x + 3$. If you want to perform the multiplication first, you write the expression $(6 \times x) + 3$; and if you want to perform the addition first, you write it $6 \times (x + 3)$.

As you know well by this time, the multiplication sign is generally omitted in algebraic expressions. The two expressions just given can be written more simply as $(6x) + 3$ when you want to indicate that the multiplication is to be performed first, or as $6(x + 3)$ when the addition is to be performed first.

As usual, though, mathematicians omit symbols when they can. In the expression $6x$, the two symbols hug each other so closely that it seems unnecessary to press them together even more closely by means of a parenthesis. The parenthesis is assumed and the expression $(6x) + 3$ is written simply $6x + 3$; just as $+5$ is usually written simply as 5, and $1x$ is written simply as x.

This makes it all the more important, however, to remember to include the parenthesis thus, $6(x + 3)$, when you want the addition performed first. If the expression is written simply $6x + 3$, it is assumed as a matter of course that the parenthesis goes (invisibly, to be sure) about the $6x$.

The same holds true for division. If you have the expression $6 \div x - 1$, you can write that either $(6 \div x) - 1$, or $6 \div (x - 1)$, depending on whether you want the division or the subtraction to be performed first. The algebraic way of writing these two expressions should be $\left(\dfrac{6}{x}\right) - 1$, or $\dfrac{6}{(x-1)}$. In

the former case, however, the 6 and the x are again so closely hugged that the parenthesis is omitted as nonessential and the expression is written simply as $\dfrac{6}{x} - 1$.

As for the expression $\dfrac{6}{(x-1)}$, the parenthesis can be dropped because the mark is extended to cover the entire expression $x - 1$ in this fashion: $\dfrac{6}{x-1}$.

Now we have our answer as to what constitutes a term. Any expression that does not include an operation, such as x or 75, is a term. In addition, any expression that includes one or more operations but is enclosed in a parenthesis is a term. Thus, $x - 1$ is an expression made up of two terms, but $(x - 1)$ is made up of but a single term.

It is important to remember that multiplications and divisions are treated as though they are enclosed in parentheses even when those parentheses are not written in. Thus, $6x$ is a single term, and $\dfrac{6}{x-1}$ is a single term.

TRANSPOSING IN ORDER

Now we can talk about equations involving both multiplication and addition, such as

$$6x + 3 = 21$$

Here we are faced with a dilemma. If the equation were merely $x + 3 = 21$, we would have no problem. Transposing would make it $x = 21 - 3$, so that we see at once that x equals 18.

If, on the other hand, the equation were simply $6x = 21$, transposition would set x equal to $\frac{21}{6}$ or $3\frac{1}{2}$.

But in the equation $6x + 3 = 21$, both multiplication and addition are involved, and so the problem arises as to which transposition to make first. If we transpose the addition first and the multiplication second, then:

$$6x + 3 = 21$$
$$6x \quad = 21 - 3 \quad \text{(or 18)}$$
$$x \quad = \frac{18}{6} \quad \text{(or 3)}$$

But if we transpose the multiplication first and the addition second:

$$6x + 3 = 21$$
$$x + 3 = \frac{21}{6} \qquad \left(\text{or } 3\frac{1}{2}\right)$$
$$x \quad = 3\frac{1}{2} - 3 \quad \left(\text{or } \frac{1}{2}\right)$$

Now if the same equation is going to yield us two different answers according to the method we use to solve it, we are faced with an inconsistency that must be removed. One or the other method must be forbidden. To decide which method to forbid, let's go back to our parentheses.

Remember that a multiplication in algebra is always treated as though it were within parentheses. The expression $6x + 3$ could be more clearly written as $(6x) + 3$, which means we must get a numerical value for $6x$ before we can add 3 to it. But we can't get a numerical value for $6x$ because we don't know what quantity x represents.

The only thing we can do, then, is to leave the expression $6x$ just as it is and to keep from breaking it apart as long as there is an operation of addition in the expression. To transpose the 6 would be to break it apart, so this can't be done.

We can make a general rule, then. When one side of an equation consists of more than one term, we can transpose only complete terms. When, however, one side of an equation consists of but a single term, portions of that term can be transposed.

If we look again at the equation

$$6x + 3 = 21$$

we see that $6x + 3$ contains two terms. The figure

3 is a whole term by itself and can be transposed. The figure 6, on the other hand, is only part of the term $6x$ and cannot be transposed as long as more than one term exists.

Therefore, we transpose the 3. That gives us $6x = 18$. Now $6x$ is the only term on its side of the equation and the 6 can be transposed. The answer is that $x = 3$, and it is the only answer. To work the equation so that x is made to be equal to $\frac{1}{2}$ breaks the rules of transposition.

If, on the other hand, we had the equation

$$6(x + 3) = 21$$

we have the single term $6(x + 3)$ on the left-hand side of the equation. Does that mean we can transpose either the 6 or the 3 at will? Well, remember that the expression includes a multiplication and therefore behaves as though it were written this way: $[6(x + 3)]$. This means, you will recall, that the operation inside the innermost parenthesis, which, in this case, is $x + 3$, must be performed first. Since this operation cannot be performed first because a literal symbol is involved, the alternative is to keep it intact the longest. In other words, if operations are performed from innermost parentheses outward, transpositions must be performed from outermost parentheses inward.

So we transpose the 6 first, that being included in the outermost parenthesis, and have

$$x + 3 = \frac{21}{6} \quad \left(\text{or } 3\frac{1}{2}\right)$$

Now, with only the innermost parenthesis left (which, however, is now omitted, because it is usual practice to omit parentheses that enclose entire expressions), we can transpose the 3:

$$x = 3\frac{1}{2} - 3 \quad \left(\text{or } \frac{1}{2}\right)$$

By remembering the two rules:
(1) When more than one term exists, transpose only entire terms;
(2) When a single term exists, perform transpositions from outermost parentheses inward;
you will always end up with the only possible solution for the equation. That, in fact, is the purpose of the rules, to make sure that only one solution is arrived at and to eliminate the possibility of wrong turnings and consequent inconsistencies.

To be sure, it may now seem to you that solving equations must become a matter of long brooding while you count terms and locate parentheses. Actually, believe it or not, this is not so. Once you become accustomed to manipulating equations, you get the hang of which transpositions come before which. The whole thing becomes so mechanical and automatic that you never give the matter a thought.

Of course, the only way in which you can arrive at such a happy state of affairs is to solve equation after equation after equation. Practice makes perfect in manipulating equations just as in manipulating a piano keyboard.

It is for this reason that school texts in algebra bombard the student with hundreds of equations to solve. It may be hard for the student to realize *why* there must be an endless drill while he is undergoing it, but that is like the finger exercises on the piano. Eventually it pays off.

And, it stands to reason, the better you understand what you are doing and why you are doing it, the more quickly it will pay off.

5

Backwards, Too!

So FAR, I have had literal symbols involved in algebraic operations on only one side of an equation. There is no reason why matters should be so restricted. Literal symbols could be present on both sides. The x's could be both here and there, so to speak.

Here's an example of the type of problem that would give you such a double-jointed situation. Suppose Jim owns a certain number of books and Bill owns twice as many. Jim buys five books to add to his supply and Bill buys only one. They end up with the same number of books. How many did each have to begin with?

I started with the statement "Jim owns a certain number of books," so let's call that "certain number" x. Bill owns twice as many, or $2x$. Jim buys five books, making his total $x + 5$; while Bill buys one, making his total $2x + 1$. They end with the same number so

$$2x + 1 = x + 5$$

This brings up a question at once. As long as I kept literal symbols on one side of the equation only, it may have seemed natural to keep that side on the left. Now we have literal symbols on both sides, so which side ought to be on the left? Might I not have written the equation this way?

$$x + 5 = 2x + 1$$

And if I had, would it make any difference?

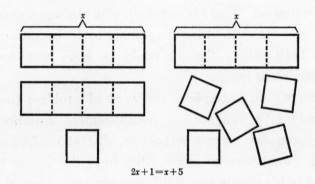

$2x + 1 = x + 5$

Perhaps this has never occurred to you as something to question. If $3 + 3 = 6$, then surely $6 = 3 + 3$. It can make no difference which way we write it, can it? Or, if we want to make it a general rule, we can say that if $a = b$, then $b = a$.

This is the sort of thing that is sometimes considered "obvious" or "self-evident." Everyone accepts it without question. Such a self-evident statement accepted by everybody is called an "axiom." I have used a number of others in this book, too. For instance, the statement I introduced in Chapter 2, that you can add the same quantity to both sides of an equation and still have it an equation, is another example of an axiom. This could be put into words as "equals added to equals are equal." It could also be put into the form of general equations, thus: if $a = b$, then $a + c = b + c$.

You may wonder why it is necessary to take any special note of axioms if everyone accepts them. Oddly enough, until nearly 1900 mathematicians were a little careless about the axioms they used, but then questions arose as to how much of the mathematical system was really justified by logic.

Men such as the Italian mathematician Giuseppe Peano (pay-AH-no) and the German mathematician David Hilbert corrected this by carefully listing all the axioms they were going to use. They then deduced all the rules and statements of algebra from those axioms only. Naturally, I don't try to do anything of the sort in this book, but you might as well know that it can be done. (Furthermore, mathematicians began to consider axioms simply as the basic beginnings for any system of orderly thinking and didn't worry any longer about whether they were "obvious" or not. Some axioms, actually, are not at all obvious.)

Now let's go back to the equation

$$2x + 1 = x + 5$$

One thing we can do is to transpose the 1 from the left-hand side of the equation to the right-hand side, exactly as we have been doing all along. But suppose that for some reason we wanted to transpose the 5 from the right-hand side to the left-hand side instead, changing the operation to the inverse, so that the equation reads

$$2x + 1 - 5 = x$$

If you remember how I first showed that transposition could be allowed in the first place (in Chapter 2), you will remember that I did it by making use of the axiom "equals added to equals are equal." This, of course, means that "equals subtracted from equals are equal," since a subtraction is only the addition of a negative number.

We can make use of that axiom here, too. Suppose we subtract 5 from each side of the original equation. We have

$$2x + 1 - 5 = x + 5 - 5$$

or

$$2x + 1 - 5 = x$$

The same axiom, you see, that allows transposition from left to right, allows it also from right to left. We can do it frontward, and we can do it backward, too!

There is no doubt, then, that we can maneuver the numerical symbols either way and collect them all on the left-hand side or on the right-hand side, whichever suits our fancy. What about the literal symbols, though? So far in the book, I have not transposed a literal symbol.

But why not, if I wish? I said very early in the book that both literal and numerical symbols represented quantities and that both were subject to the same rules and could be treated in the same way. If numerical symbols such as 1 or 5 can be transposed, then a literal symbol such as x can be transposed and that's that.

Therefore, in the equation

$$2x + 1 = x + 5$$

let's transpose the 1 from left to right in the usual manner and the x from right to left in the backward manner so that we have all the literal symbols on the left side and all the numerical symbols on the right side, thus:

$$2x - x = 5 - 1$$

In both cases, the operation was changed to the inverse, so that both additions became subtractions.

The equation now works out at a glance to

$$x = 4$$

and that is the solution of our problem.

What we have decided is that Jim owned 4 books and that Bill owned twice as many, or 8. When Jim bought 5 more books and Bill bought 1 more book, they ended, exactly as the problem stated, with an equal number of books, 9 apiece.

Does it strike you that this is a long and complicated way of solving the problem? Not at all. It only seems long and complicated because I am taking the trouble of explaining each step in detail. Once you have the system of algebraic manipulation down pat, however, you can go through such equations like a streak. In fact, an equation as simple as the one with which I have been working in this chapter would be so little trouble to you that you could solve it in your head in short order.

CONSISTENCY AGAIN

Yet for all that the equation is so simple, I am not through extracting the juice from it even now. Let's see if reversing the direction of transposition might not involve us in an inconsistency after all.

Here's the equation one more time:

$$2x + 1 = x + 5$$

I have just solved for x by transposing in such a way as to get all the literal symbols on the left-hand side and all the numerical symbols on the right-hand side. Is there some reason we are forced to do this? Or would matters have gone as well if we had transposed all the literal symbols to the right-hand side, rather than the left; and all the numerical symbols to the left-hand side, rather than the right?

The straightforward thing is to try it and see if we get the same answer when working it backward.

We therefore transpose the 5 from right to left and the $2x$ from left to right. (The expression $2x$ is a single term, remember, and must be transposed intact as long as it forms part of a polynomial.) The result is

$$1 - 5 = x - 2x$$

which works out to

$$-4 = -x$$

Now there is room for a little doubt. When we had transposed terms in the forward direction, we ended with $x = 4$. When we transposed terms in the backward direction, we ended with $-4 = -x$. Are these different answers? Have we uncovered an inconsistency?

To check that, let's remember that in Chapter 3

I said that both sides of an equation could be multiplied by the same number without spoiling the equation. (This is another axiom, which can be expressed as "equals multiplied by equals are equal," or, if $a = b$, then $ac = bc$.)

Suppose, then, that we take the expression $-4 = -x$ and multiply each side of the equation by -1. This would give us

$$(-1)(-4) = (-1)(-x)$$

By following the rule of multiplication of signs, this becomes

$$4 = x$$

And by the axiom which tells us that it doesn't affect the equation if we interchange the right and left sides, we can say this is equal to

$$x = 4$$

So you see, we come out with the same answer after all, no matter in which direction we make our transpositions. From now on, we can certainly feel secure in transposing either forward or backward.

And, incidentally, the trick of multiplying by -1 can be used to change all signs on both sides of any equation. Using general symbols, we can say:

If $\qquad a + b = c + d$

then $\qquad -a - b = -c - d$

And if $\qquad a - b = c - d$

then $\qquad -a + b = -c + d$

Since an expression such as $-a + b$ looks more familiar to us if written $b - a$, the last set of expressions might be written:

If $\qquad a - b = c - d$

then $\qquad b - a = d - c$

As I continue to heap up the rules of manipulating equations and show how flexible they are, you may be getting the idea that they are a wonderfully mechanical way of getting the truth out of a problem. So they are, but don't expect too much out of the situation. We have gone far enough now for me to be able to explain that algebraic manipulation cannot get any more truth out of an equation than is put into it in the first place.

Suppose that, instead of the equation $2x + 1 = x + 5$, which we have been pounding from every side in this chapter, I were to present you with the very similar equation

$$x + 1 = x + 5$$

By transposing, you get

$$x - x = 5 - 1$$

or

$$0 = 4$$

which is a nonsensical answer.

How did that happen? The algebraic manipulation was strictly according to rule. Are the rules wrong, then?

Well, look at the equation $x + 1 = x + 5$. This says that if you take a particular number and add 1 to it, you get a result which is the same as that obtained when you add 5 to it.

But this is nonsensical. Any number must yield two different sums if two different quantities are added to it. (This can be stated as "unequals added to equals are unequals," or, if $a \neq b$, then $x + a \neq x + b$.)

Therefore $x + 1$ cannot equal $x + 5$ for any value of x at all, and $x + 1 = x + 5$ is a false equation. To pretend that it is a true equation and to use rules of manipulation that are only intended for true equations does us no good. We start with nonsense and we end with nonsense.

Always be sure, then, that you are making sense in the first place and the rules of algebra will then take care of you. If you're not making sense to begin with, then nothing can take care of you, algebra least of all.

6

The Matter of Division

OF THE FOUR algebraic operations I have discussed so far in the book, division certainly seems the one hardest to handle. If you begin with whole numbers and confine yourself only to addition and multiplication, you always end up with whole numbers. If you deal with subtraction, you have to add negative numbers to the list, but they are still whole numbers.

If, however, you subject whole numbers to division, you more often than not end up with fractions, which are harder to handle than whole numbers are. And if you try to convert fractions to decimals by means of further division, you may find yourself with an endless decimal. (Try to divide 10 by 7 in order to get a decimal value and see for yourself.)

It is not surprising, then, that algebraic equations that involve division are sometimes a touch more complex than are those that do not.

Often, an equation involving division (and always remember we can never divide by zero) can be

treated in just the same way as were those involving
multiplication which I described in the previous
chapter. In the equation

$$\frac{x}{4} + 3 = 11$$

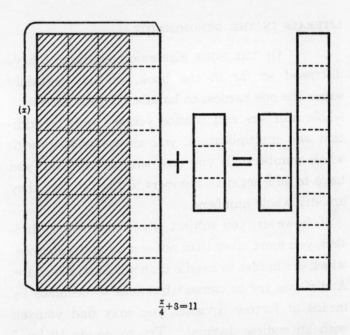

$$\frac{x}{4} + 3 = 11$$

the $\frac{x}{4}$ is a single term and is treated as though it
were enclosed in a parenthesis. We don't break it
up, therefore, but transpose the 3:

$$\frac{x}{4} = 11 - 3 \qquad \text{(or 8)}$$

It is only then that we can transpose the 4, changing the operation from division to its inverse, multiplication, of course:

$$x = (8)(4) \qquad \text{(or 32)}$$

On the other hand, if the equation were

$$(x + 3)/4 = 11$$

then the entire division, still being taken as enclosed in a parenthesis, becomes a parenthesis within a parenthesis: $[(x + 3)/4]$.

The rule is that we break up parentheses from the outside inward. We break up the outer parenthesis first, then, by transposing the 4 and leaving the inner $(x + 3)$ intact:

$$x + 3 = (11)(4) \qquad \text{(or 44)}$$

Then, and only then, can we transpose the 3:

$$x = 44 - 3 \qquad \text{(or 41)}$$

An additional touch of complexity, however, arises from the fact that most often the division is represented by a horizontal line rather than a shilling mark, thus: $\dfrac{x + 3}{4}$. Here both parentheses are omitted, yet the expression must still be treated as though it were $\left[\dfrac{(x + 3)}{4}\right]$, even though no paren-

theses are actually visible. You would still have to
remember to transpose the 4 first, as part of the
outer parenthesis, and then the 3 as part of the
inner parenthesis.

But don't be downhearted. Only sufficient drill
is required to make it all come second nature so
that you will never give a thought to omitted
parentheses. And if you know what you are doing
to begin with, you will find that you won't even
need very much practice to achieve this happy
result.

A more serious touch of complexity arises from
the fact that a literal symbol might easily be in the
denominator of a fraction. This is something we
haven't considered before, but there is no reason
why it can't come up. Here is an example:*

$$\frac{10}{x-5} = 5$$

The expression on the left is a parenthesis within
a parenthesis, $\left[\frac{10}{(x-5)}\right]$, and we ought, therefore,
to transpose the 10 first as being part of the outer

---

* In this example, you know at once that whatever x
may be equal to, it cannot be equal to 5. Do you see
why? If x were equal to 5, then $x - 5$ would be equal
to zero, and you would have the expression $\frac{10}{0}$. This
involves division by zero, which is *forbidden*.

parenthesis. So far, though, we have never transposed the numerator of a fraction. How ought we
to go about it?

One way of managing would be to convert the
division into a multiplication. Since $\dfrac{10}{x-5}$ represents 10 divided by $(x-5)$, we can write it instead
as 10 multiplied by the reciprocal of $(x-5)$, or
$(10)\left(\dfrac{1}{x-5}\right)$. For this reason we can change the
original equation to read

$$(10)\left(\frac{1}{x-5}\right) = 5$$

and now we can transpose the 10 in the usual
manner so that

$$\frac{1}{x-5} = \frac{5}{10}$$

Having successfully transposed the numerator, however, we find we are still left with a fraction on
the left, and one that contains the literal symbol
in the denominator even yet.

An alternative plan of attack would have been to
transpose the denominator as a whole, parenthesis
and all. The presence of the inner parenthesis in
$\left[\dfrac{10}{(x-5)}\right]$ tells us we can't split up the expression
$(x-5)$ while the outer parenthesis exists. How-

ever, there is nothing in the rules to prevent us from transposing it intact, *without* splitting it up.

In this way, if

$$\frac{10}{x-5} = 5$$

then

$$10 = 5(x-5)$$

Transposing backward,

$$\frac{10}{5} = x - 5$$

and

$$\frac{10}{5} + 5 = x$$

so that x is equal to 7, which is the solution of the problem.

Still another method of tackling the equation would be to sidestep the problem of the literal symbol in the denominator altogether. It is easy to put the literal symbol into the numerator by just taking the reciprocal of the fraction and thus turning it upside down. Instead of writing $\frac{10}{x-5}$, we would write $\frac{x-5}{10}$.

Ah, but what does that do to the equation?

Well, let's go back to numerical symbols. If $\frac{10}{5} = 2$, what happens if we convert that fraction into its reciprocal $\frac{5}{10}$? You can see that $\frac{5}{10}$ is equal to $\frac{1}{2}$, which is the reciprocal of 2.

In other words (and you can check this by trying other examples), an equation remains intact if you take the reciprocal of *both* sides of the equation. In general expressions:

If $\qquad \frac{a}{b} = \frac{c}{d}$ (where neither a, b, c, nor d may equal zero)

then $\qquad \frac{b}{a} = \frac{d}{c}$

The equation

$$\frac{10}{x - 5} = 5$$

can therefore be converted to

$$\frac{x - 5}{10} = \frac{1}{5}$$

and by proper transpositions

$$x - 5 = \left(\frac{1}{5}\right)(10) \qquad \text{(or 2)}$$

and

$$x = 2 + 5 \qquad \text{(or 7)}$$

The solution is, as before, that x is equal to 7.

THE PROBLEM OF FRACTIONS

This rule of reciprocals, however, exposes the beginner to a pitfall into which it is all too easy to fall — a very attractive but very deadly trap. Suppose you have the equation

$$\frac{1}{2x} + \frac{1}{3x} = \frac{1}{5}$$

Remembering the rule of reciprocals, you might try to take the reciprocals of all the fractions concerned, changing the equation to read $2x + 3x = 5$, which can at once be changed to $5x = 5$, and by transposition to $x = \frac{5}{5}$ or 1.

But if you substitute 1 for the x of the original equation, you find that $\frac{1}{(2)(1)} + \frac{1}{(3)(1)} = \frac{1}{5}$, or $\frac{1}{2} + \frac{1}{3} = \frac{1}{5}$, which is not a true equation since, in actual fact, $\frac{1}{2} + \frac{1}{3} = \frac{5}{6}$.

What is wrong?

If you tried by experimenting with different equations, you would find that the rule of reciprocals works only when the reciprocal of each side of an equation is taken as a whole, and not as separate parts. In other words if each side of the equation is a monomial, the reciprocal of each side of the equation can be taken and the equation will stay intact. If one or both sides of the equation is a polynomial, however, taking the reciprocal of each term separately almost inevitably reduces the equation to nonsense.

Then what if you *do* have a binomial to deal with, as on the left side of the equation

$$\frac{1}{2x} + \frac{1}{3x} = \frac{1}{5}$$

Clearly, what must be done, if you want to apply the rule of reciprocals, is to convert that binomial into a monomial, a single term. And that means that we are faced with the problem of the addition of fractions.

This subject, fortunately, comes up in ordinary arithmetic and is dealt with thoroughly there. The same rules developed in arithmetic can be used in algebra, so a quick review is all that is needed.

To begin with, there is no problem in adding fractions that have the same denominator. When that happens, all we have to do is add numerators and leave the denominator as is. Thus, $\frac{5}{3}$ plus $\frac{2}{3}$ equals $\frac{7}{3}$, and $\frac{16}{19}$ plus $\frac{54}{19}$ equals $\frac{70}{19}$.

In adding fractions with different denominators, you must somehow change them into fractions with the same denominators or you are stymied. And this must be done without changing the value of the fractions, of course.

Now any fraction can be changed in form without change in value if the numerator and denominator are multiplied by the same number. Thus, if you multiply the fraction $\frac{1}{2}$, top and bottom, by 2, you end with $\frac{2}{4}$. If you multiply it top and bottom by 3, you have $\frac{3}{6}$; if you multiply it top and bottom by 15, you have $\frac{15}{30}$. The value doesn't change, you see, even though the form does, for you know from your arithmetic that $\frac{2}{4}$ equals $\frac{1}{2}$, and that $\frac{3}{6}$ and $\frac{15}{30}$ also equal $\frac{1}{2}$.

Now let's look at the equation that gave us trouble at the beginning of this section:

$$\frac{1}{2x} + \frac{1}{3x} = \frac{1}{5}$$

If we multiply the denominator of the fraction $\frac{1}{2x}$ by 3 and the denominator of the fraction $\frac{1}{3x}$ by 2, we end with the same denominator in each case, $6x$. That is fine, so far, but if we multiply a denominator by 2, we must also multiply the numerator of that fraction by 2 to keep the value unchanged; and if we multiply another denominator by 3, we must also multiply the corresponding numerator by 3. The equation must therefore be written:

$$\frac{(1)(3)}{(2x)(3)} + \frac{(1)(2)}{(3x)(2)} = \frac{1}{5}$$

or

$$\frac{3}{6x} + \frac{2}{6x} = \frac{1}{5}$$

Fractions with the same denominator are added together easily, so that

$$\frac{5}{6x} = \frac{1}{5}$$

Now, at last, you can apply the rule of reciprocals, and write

$$\frac{6x}{5} = 5$$

so that, by transposition,

$$6x = (5)(5) \qquad \text{(or 25)}$$

and

$$x = \frac{25}{6}$$

That this is the correct answer can be shown if we substitute $\dfrac{25}{6}$ for x in the original equation. That would look this way:

$$\frac{1}{2\left(\dfrac{25}{6}\right)} + \frac{1}{3\left(\dfrac{25}{6}\right)} = \frac{1}{5}$$

or, by ordinary arithmetic,

$$\frac{1}{\dfrac{25}{3}} + \frac{1}{\dfrac{25}{2}} = \frac{1}{5}$$

Now $\dfrac{1}{\dfrac{25}{3}}$ represents 1 divided by $\dfrac{25}{3}$ which can be also represented as 1 multiplied by the reciprocal of $\dfrac{25}{3}$, or $(1)\left(\dfrac{3}{25}\right)$, which is equal to $\dfrac{3}{25}$. By the same reasoning, we can show that $\dfrac{1}{\dfrac{25}{2}}$ is equal to $\dfrac{2}{25}$. In fact, we can make the general rule that

$$\frac{1}{\dfrac{b}{a}} \equiv \frac{a}{b}$$

so that our equation can be written

$$\frac{3}{25} + \frac{2}{25} = \frac{1}{5}$$

or

$$\frac{5}{25} = \frac{1}{5}$$

which is, of course, a true equation, and shows that we located the correct solution that time.

7

The Ins and Outs of Parentheses

I AM SURE that by now you are perfectly satisfied that parentheses can be useful in helping solve an equation without confusion. Yet there are times when parentheses are a positive embarrassment. I will give you an example of this.

Imagine two rectangles of known height, both being 5 inches high. However, you don't know the widths exactly; all you know is that one rectangle is 3 inches wider than the other. You also know that the total area of the two rectangles, taken together, is 35 square inches. Now the question is: What are the widths of the rectangles?

In order to determine the area of a rectangle, it is necessary to multiply the width by the height. In other words, a rectangle that is 17 inches wide and 12 inches high is 17 times 12, or 204 square inches* in area.

---

* I could stop here and discuss the fact that the length and width of a square are measured in inches while the area must be measured in square inches, but that would take me far from the main subject of the book.

With that understood, let's tackle the problem and begin by calling the width of the narrower rectangle x inches. The other rectangle, which is 3 inches wider, would naturally be $x + 3$ inches wide. In each case, the width must be multiplied by the height (5 inches) to obtain the area. Therefore the area of the narrower rectangle is $5x$ square inches, while that of the other is $5(x + 3)$ square inches.

Since the sum of the areas is 35 square inches, we can write

$$5(x + 3) + 5x = 35$$

(We don't have to write "inches" and "square inches" in such an equation, because in ordinary algebra we are dealing only with the quantities. However, we must always keep in mind the correct "units of measurement.")

In an equation such as this, how do we solve for x? Our natural impulse is to get x all by itself on one side of the equation and all the numerical symbols on the other, but how can we do that with the parenthesis barring the way?

We can transpose the $5x$ and then the 5 and then

If you are not well acquainted with this sort of thing and would like to go into it a bit more when you have a chance, you will find it (and other matters involving measurement) discussed in my book *Realm of Measure* (Houghton Mifflin, 1960).

the 3, so the equation will take up the following forms:

$$5(x + 3) = 35 - 5x$$

$$x + 3 = \frac{35 - 5x}{5}$$

$$x = \frac{35 - 5x}{5} - 3$$

Now we have an x all by itself on the left-hand side of the equation, but alas, we also have a literal symbol on the right-hand side. In order to get the $5x$ back on the left, we have to transpose first the 3, then the 5, and only then the $5x$, and we are back where we started.

No, if we are to get anywhere we must get rid of the parenthesis which keeps us from combining the x within it and the $5x$ outside it into a single term. How do we do that though?

Let's take a close look at the term $5(x + 3)$ and ask ourselves what we would do if only numerical symbols were involved. Suppose we had the term $5(2 + 3)$ instead. Of course, this is no problem since $2 + 3$ equals 5 so that we can change the expression to $(5)(5)$ and come out with an answer of 25 at once.

However, we can't combine x and 3, as we can

combine 2 and 3, so we must ask ourselves: How can we work out the value of the expression $5(2 + 3)$ if, for some reason, we are forbidden to combine the 2 and the 3?

The natural thing to try, I think, is the multiplication of each number within the parenthesis by 5. This gives us $(5)(2)$ or 10, and $(5)(3)$ or 15, and come to think of it, $10 + 15$ is 25. You can try this with any combination of numbers and you will find it will work. Thus $6(10 + 5 + 1)$ is $(6)(16)$ or 96. But if you multiply the 6 by each number within the parenthesis separately, you have $(6)(10)$ or 60, $(6)(5)$ or 30, and $(6)(1)$ or 6; and if you add the products, you have $60 + 30 + 6$, which also comes to 96.

Using general symbols, we can say that

$$a(b + c) = ab + ac$$

Certainly, then, this means that we can change

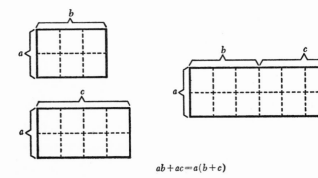

$$ab + ac = a(b + c)$$

$5(x + 3)$ to $(5)(x) + (5)(3)$ or $5x + 15$. The original equation would now read

$$(5x + 15) + 5x = 35$$

I have retained the parenthesis just to show you I haven't forgotten it is there, but since only additions are involved and the order of addition doesn't matter, we might as well drop it:

$$5x + 15 + 5x = 35$$

We are now free to add the two x-containing terms and to proceed with transpositions:

$$10x + 15 = 35$$

$$10x = 35 - 15 \qquad \text{(or 20)}$$

$$x = 20/10 \qquad \text{(or 2)}$$

If we substitute 2 for x in the original equation, we have

$$5(2 + 3) + (5)(2) = 35$$

$$25 + \quad 10 \quad = 35$$

and, as you see, all is well.

Of the rectangles I spoke of at the beginning of the chapter, then, one is 2 inches wide and the other (which is 3 inches wider) is 5 inches wide. The area of the first is 2 times 5 or 10 square inches; that of the second is 5 times 5 or 25 square inches; so that the total area is indeed 35 square inches as the problem specified.

SUBTRACTING AND PARENTHESES

Now we come to another possible pitfall. In the previous section, I removed a parenthesis by stating quickly that only additions were involved. Is this really justified? Let's try it in an expression using numerical symbols only, such as $7 + 4(2 + 6)$. To work out the value we first combine the numbers within the parenthesis to give 8 so that the expression reads $7 + (4)(8)$. Multiplying first, we have $7 + 32$ or 39.

Let's begin again and remove the parenthesis, so that we have $7 + (4)(2) + (4)(6)$, or $7 + 8 + 24$. The answer is still 39, you see.

So far, good; but now let's just make a slight change and consider the following expression: $7 - 4(2 + 6)$. By combining the numbers within the parenthesis first, we have $7 - (4)(8)$, or $7 - 32$, or -25 as the value of the expression.

However, if we remove the parenthesis exactly as before, we have $7 - (4)(2) + (4)(6)$, or $7 - 8 + 24$, or $+23$. Now we are faced with an inconsistency, for there are two answers, -25 and $+23$. What is wrong?

Obviously the minus sign has introduced a complication. Therefore, let's remove it and change it to a plus sign. This can be done, for you remember that we decided long ago that $a - b$ can be written $a + (-b)$.

Instead of writing the expression, then, as $7 - 4(2 + 6)$, let's write it $7 + (-4)(2 + 6)$. Now do you see what we've done? In removing the parentheses we are going to have to multiply each number inside the parenthesis not by 4, but by -4, and that means a change in signs.

The expression becomes $7 + (-4)(2) + (-4)(6)$, or $7 + (-8) + (-24)$. And since $a + (-b)$ can be written $a - b$, the expression can be written $7 - 8 - 24$, or -25.

Now observe carefully. When you began with the expression $7 + 4(2 + 6)$, you ended, after removing the parenthesis, with $7 + 8 + 24$. But when you began with the expression $7 - 4(2 + 6)$, you ended, after removing the parenthesis, with $7 - 8 - 24$. The signs inside the parenthesis had been changed!

This is true even when a parenthetical expression is simply subtracted, with no number visible outside it, as in the case of $5 - (2 + 3)$. If you combine the numbers inside the parenthesis first, the value of the expression is $5 - 5$, or 0. If, however, you try simply to drop the parenthesis, you have $5 - 2 + 3$, which comes out to 6. Instead, consider such an expression as $5 - 1(2 + 3)$, which is what it really is, since 1 times any expression is equivalent to the expression itself.

This can be written $5 + (-1)(2 + 3)$, and, in removing parentheses, we have $5 + (-1)(2) + (-1)(3)$, or $5 + (-2) + (-3)$, or $5 - 2 - 3$, which comes out to zero, as it should.

We can set up general rules as follows:

$$a - b(c + d) \equiv a - bc - bd$$

$$a - b(c - d) \equiv a - bc + bd$$

$$a - b(-c - d) \equiv a + bc + bd$$

and so on.

Generally, this is stated to the effect that when a minus sign appears before a parenthesis, all the positive signs within it must be changed to negative and all the negative signs to positive when the parenthesis is removed. The student is drilled endlessly to make sure he learns to do this automatically. Now that you see why the reversal of sign must be made, you should have very little trouble remembering to do it — I hope.

BREAKING UP FRACTIONS

Of course, don't get the idea that parentheses must always be removed at all costs the instant you see one. Sometimes, as you know, having them there helps. In fact, there are actually times when you can make things simpler for yourself by putting parentheses into an equation where none existed before.

To see how that can come about let's begin by looking at the fraction $\frac{2}{10}$.

Since the value of a fraction is not altered if both numerator and denominator are divided by the same quantity, $\frac{2}{10}$ can be divided, top and bottom, by 2, and the fraction $\frac{1}{5}$ is obtained. From ordinary arithmetic this has probably become second nature to you so that you know at a glance that $\frac{2}{10}$ is equal to $\frac{1}{5}$, that $\frac{3}{9}$ is equal to $\frac{1}{3}$, that $\frac{14}{21}$ is equal to $\frac{2}{3}$, $\frac{25}{65}$ to $\frac{5}{13}$, and so on.

When, by dividing a fraction, top and bottom, you reach the smallest possible combination of whole numbers which retain the value, that fraction has been "reduced to lowest terms." In other words, $\frac{14}{21}$ has not been reduced to lowest terms, but the equivalent fraction, $\frac{2}{3}$, has.

Now let's be a little more systematic about

reducing fractions to lowest terms. A number may be broken into two or more smaller numbers which, when multiplied together, give the original number as a product. The smaller numbers so obtained are called "factors" and when the original number is expressed as a product of the smaller numbers it is said to be "factored."

For instance, you can factor 10 by writing it as (5)(2). Consequently, 5 and 2 are both factors of 10. You can factor 12 as (4)(3), or (2)(6), or (2)(2)(3). Sometimes it is even convenient to write 10 as (1)(10) or 12 as (1)(12).*

Suppose now that the numerator and denominator of a fraction are both factored and it turns out that at least one factor in the numerator is equal to one factor in the denominator. For instance, the fraction $\frac{14}{21}$ can be written as $\frac{(7)(2)}{(7)(3)}$. If you divide

---

* In many cases, a number can only be expressed as itself times 1, with no other form of factoring possible. For instance, 5 can be written as (5)(1), 13 as (13)(1), 17 as (17)(1), and in no other way. Such numbers are called "prime numbers." In factoring a "composite number," one that isn't prime, it is often convenient to factor it to prime numbers. For instance, 210 can be factored as (10)(21), or as (70)(3), or in any of a number of other ways, but if you work it down to prime numbers, 210 is equal to (2)(3)(5)(7). It is an important theorem in that branch of mathematics known as "theory of numbers" that every composite number can be broken down to prime factors in only one way.

both numerator and denominator by 7, you have

$$\frac{\left(\frac{7}{7}\right)(2)}{\left(\frac{7}{7}\right)(3)}$$, which is equal to $\frac{(1)(2)}{(1)(3)}$ or $\frac{2}{3}$. Rather

than go through this, the student quickly learns that all he needs to do is cross out any factor that appears both top and bottom. The 7's are "canceled" and the result is that the fraction is reduced to lowest terms, as you learned to do in arithmetic.

Of course, it is important to remember that the canceling is only a short cut and that what you are really doing is dividing the numerator and the denominator by equal numbers. Canceling numbers wildly can lead to many a pitfall and it will never hurt, in doubtful cases, to go back to dividing both parts of a fraction to make sure that you are doing the right thing.

When numerator and denominator have no factor in common other than 1, nothing can be canceled and the fraction is already at its lowest terms. Thus, the fraction $\frac{15}{22}$ can be written as $\frac{(3)(5)}{(2)(11)}$. There is no opportunity for cancellation here and the fraction is at lowest terms.

Nevertheless, before you can be certain that a fraction *is* at lowest terms, you must be sure you have factored the numerator and denominator as far as you can. Suppose you have the fraction $\frac{24}{30}$.

Since 24 can be factored as (8)(3) and 30 can be factored as (6)(5), you could write the equation as $\frac{(8)(3)}{(6)(5)}$. There would seem to be no factors in common — but wait. After all, 8 can be factored as (2)(2)(2) and 6 can be factored as (2)(3). Instead of writing 24 as (8)(3), let's write it as (2)(2)(2)(3) and instead of writing 30 as (6)(5), let's write it as (2)(3)(5). Now the fraction can be written as $\frac{(2)(2)(2)(3)}{(2)(3)(5)}$ and we can cancel the 3 and one of the 2's. The fraction then becomes $\frac{(2)(2)}{5}$, or $\frac{4}{5}$. With no more factors in common, the fraction is in lowest terms.

Now let me warn you against a pitfall. Once a student has learned to cancel, he is usually so eager to do so that, as often as not, he will do so where the rules don't permit it. Remember that cancellation involves factors and that factors are themselves involved in multiplication. You can break up a number into two smaller numbers that will give the original number through addition as, for instance, 13 can be broken up into 7 + 6. However, 7 and 6 are *not* factors of 13 and cannot be involved in cancellation.

Suppose that you are faced with the fraction $\frac{13}{21}$.
Numerator and denominator, here, have no factors
in common other than 1 and the fraction is at lowest
terms. The eager student, however, might break it
up this way, $\frac{7+6}{(7)(3)}$, and try to cancel 7's in order
to give the result $\frac{6}{3}$ or 2, which is clearly false. Of
course, the student would see that at once and
realize something was wrong. Where more complicated expressions are involved, he would not see
the error at once and it might take him quite a
while to spot it.

INSERTING THE PARENTHESES

The rules of factoring and cancellation can be
applied to literal symbols with hardly a hitch.

For instance, suppose you have the fraction $\frac{4x}{10}$.
The 10 presents no problem; it can be factored as
(2)(5). What about the $4x$, however? How can
that be factored? Well, can't $4x$ be written as
(2)(2x)?

Let's write the fraction, then, as $\frac{(2)(2x)}{(2)(5)}$ and cancel
the 2's. The fraction, reduced to lowest terms, is
therefore $\frac{2x}{5}$.

But suppose we had started with $\frac{2x}{10}$. The 10 can still be factored as $(2)(5)$ and the $2x$ can be factored as $(2)(x)$. The fraction can be written as $\frac{(2)(x)}{(2)(5)}$, the 2's are canceled, and the fraction, in lowest terms, is $\frac{x}{5}$.

Naturally, I hope I don't have to explain why the fraction $\frac{(x + 2)}{10}$ can't be simplified by cancellation. You might write the fraction $\frac{x + 2}{(5)(2)}$ but that won't help you. The 2 in the numerator is not a factor.

On the other hand, the fraction $\frac{3(x + 2)}{15}$ can be written as $\frac{(3)(x + 2)}{(3)(5)}$ and the 3 in the numerator *is* a factor, although the 2 is not. The 3's can be canceled and the equation can be written in its lowest terms as $\frac{(x + 2)}{5}$.

Of course, it is easy to see the factors in an expression involving literal symbols when the factors happen to be right in the open. You can see that the factors of $5x$ are 5 and x and that the factors of $3(x + 2)$ are 3 and $x + 2$, but can you see the factors present in an expression like $5x + 15$?

You will, in a moment.

Earlier in the chapter, we decided that $a(b + c)$ could be written as $ab + ac$. This procedure works in reverse as well. If you start with $ab + ac$, you can write it as $a(b + c)$.

Let's take a closer look. The expression ab can be factored as $(a)(b)$, while ac can be factored as $(a)(c)$. What we have done, then, is to find the common factor, a, and place it outside a parenthesis we create for what remains.

If we go back to $5x + 15$, we see that $5x$ can be written as $(5)(x)$ and 15 as $(5)(3)$. Since 5 is the common factor, we can put it outside a parenthesis enclosing the rest and have $5(x + 3)$. If the parenthesis is now removed by the rules described at the beginning of the chapter, we get $5x + 15$ back again and consistency is upheld.

This makes it possible to simplify the fraction $\dfrac{5x + 15}{20}$.

As it stands, there seems to be nothing to cancel, for the numerator is not divided into factors. However, if the numerator is written as $5(x + 3)$, it has been factored, the two factors being 5 and $x + 3$. The denominator can, of course be factored as $(5)(4)$. The fraction can therefore be written $\dfrac{5(x + 3)}{(5)(4)}$, the 5's can be canceled and the fraction reduced to lowest terms as $\dfrac{(x+3)}{4}$.

As I promised at the beginning of the chapter, we have actually introduced a parenthesis in order to simplify an expression.

It is possible, of course, not to bother putting in the parenthesis. Each term in the numerator can be factored separately so that the fraction may be written $\dfrac{(5)(x)+(5)(3)}{(5)(4)}$. Then it is simply necessary to cancel the 5 that occurs in all the terms above and below to get $\dfrac{(x+3)}{4}$. This method is a little quicker for the person who has had enough drill at factoring, but it works only if a particular factor does indeed occur in all the terms, above and below, without exception. If even one term, either in the numerator or in the denominator, lacks the factor, canceling cannot take place. The beginner is very apt to forget this and cancel when he ought not to. The safe thing to do is to take the trouble to insert the parenthesis and draw out the common factor where you can see it plainly and know for certain that it is a common factor. Then, in peace and security, you may cancel.

Canceling, by the way, is not restricted to numerical symbols alone. Literal symbols can be treated in precisely the same way. In the fraction $\dfrac{5x}{6x}$, the x

is a factor both above and below and can be canceled

to yield the simpler fraction $\frac{5}{6}$. Or consider the

fraction $\frac{21x}{14x}$. The numerator can be factored as

$(3)(7x)$ and the denominator as $(2)(7x)$. The frac-

tion can therefore be written as $\frac{(3)(7x)}{(2)(7x)}$, and by

canceling $7x$, it is reduced to lowest terms as $\frac{3}{2}$.

However, you must always make sure in such cases
that you are not canceling an expression that happens to equal zero. Remember that cancellation
is really division and you must not divide by zero.

CANCELLATION EXTENDED

The rules of factoring and canceling, which I have
been using for fractions, will work for equations,
too. To see why this should be so, consider a
general equation in which we will call the left side L
and the right side R. The equation would be

$$L = R$$

I told you quite early in the book that both sides
of an equation can be divided by the same quantity
without spoiling the equation, so let's divide both
sides by the right side. The equation (provided, of
course, R is not equal to zero) becomes

$$\frac{L}{R} = \frac{R}{R}$$

or

$$\frac{L}{R} = 1$$

The two sides of any equation can thus be written as the numerator and denominator of a fraction equal to one. It is not surprising, then, that the rules of factoring and cancellation that apply to fractions also apply to equations.

For a specific example, consider the equation

$$10x = 15$$

By transposing, we get

$$x = \frac{15}{10}$$

We can factor the fraction in order to reduce it to lowest terms, thus:

$$x = \frac{(5)(3)}{(5)(2)} = \frac{3}{2}$$

Instead of doing this, we could factor and cancel in the equation itself to begin with. We can write $10x$ as $(5)(2x)$ and 15 as $(5)(3)$. This means we can write the equation as

$$(5)(2x) = (5)(3)$$

Now we cancel out the common factors, left and

right, just as we would do it, top and bottom, in a fraction. The equation becomes

$$2x = 3$$

and, by transposing,

$$x = \frac{3}{2}$$

The same answer is obtained, you see, whether you transpose first and factor afterward, as I did in the first case, or factor the equation first and transpose afterward, as I did in the second. In the simple equations I use in this book (just to explain the techniques of algebra) there isn't even any difference in convenience. In more complicated equations, however, it is usually far more convenient to factor and cancel as much as possible before you do anything in the way of transposition, so it is best to get used to factoring first and transposing afterward.

Naturally, you must watch out for the same pitfalls in factoring the expressions that make up equations as in factoring fractions. In the equation

$$x + 6 = 15$$

you might write it thus:

$$x + (3)(2) = (3)(5)$$

The equation is still correct, but now you will be subjected to the temptation of canceling the 3's right and left to give yourself the equation $x + 2 = 5$. Then, by transposing, you will decide that x is equal to $5 - 2$; that is, to 3. But something is wrong, for if you substitute 3 for x in the original equation, you have $3 + 6 = 15$, which is nonsense.

The mistake was in canceling the 3's, for in the expression $x + (3)(2)$, 3 is *not* a factor of the entire expression; it is a factor of one term only. Cancellation can only proceed when each side of the equation has a common factor; each side, as a whole.

When factoring an expression involving more than one term, the safest procedure is to bring in a parenthesis. Suppose, for instance, we had the equation

$$10x + 35 = 15$$

We could factor each term as follows:

$$(5)(2x) + (5)(7) = (5)(3)$$

The left side of the equation has two terms with the common factor 5. That factor can therefore be brought outside a parenthesis that encloses the remaining factors so that the equation becomes

$$5(2x + 7) = (5)(3)$$

Now the 5 on the left is indeed a factor of that entire side of the equation and not of one term only. The 5's can be canceled therefore, left and right, and the equation reads

$$2x + 7 = 3$$

Whether you solve for x in the equation $10x + 35 = 15$ or, after factoring and canceling, in the equation $2x + 7 = 3$, you will come out with the same answer. In the first case:

$$x = \frac{(15 - 35)}{10} \qquad (\text{or } -2)$$

In the second case:

$$x = \frac{(3 - 7)}{2} \qquad (\text{or } -2)$$

In other words, factoring and canceling before transposing (when properly done) does not introduce inconsistencies.

8

The Final Operations

SQUARES AND CUBES

WHAT HAVE WE done with literal symbols so far? We have combined them with numerical symbols by way of addition $(x + 7)$, subtraction $(x - 7)$, multiplication $(7x)$, and division $\left(\dfrac{x}{7}\right)$. We have even combined them with other literal symbols by means of addition $(x + 7x)$, subtraction $(x - 7x)$, and division $\left(\dfrac{x}{7x}\right)$.

The one combination I haven't used is that of a literal symbol multiplied by another literal symbol, and it is time to tackle that very situation now.

Here's how a case of multiplication among literal symbols can arise naturally in mathematics. One of the most familiar geometrical figures is the square. It is a four-sided figure with all the angles right angles,* which makes it a kind of rectangle. A

---

* A right angle is the angle formed when a perfectly horizontal line meets a perfectly vertical one.

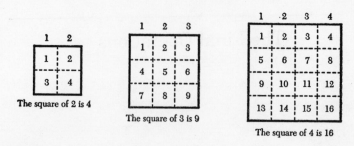

The square of 2 is 4

The square of 3 is 9

The square of 4 is 16

square differs from ordinary rectangles, though, in that all four of its sides are equal in length.

To obtain the area of a square, we must multiply the length by the height, as in any other rectangle. However, since all the sides of a square are equal, the length of a square is always equal to its height. A square that is 2 inches long is also 2 inches high; one that is 5 inches long is also 5 inches high, and so on. The area of a square with a side of 2 inches is therefore 2 times 2 or 4 square inches. The area of a square with a side of 5 inches is 5 times 5 or 25 square inches, and so on.

Because of this connection with the square, 4 is said to be the square of 2, and 25 is the square of 5. In fact, the product of any number multiplied by itself is the square of that number.

Suppose, though, we didn't know the length of the side of a particular square. We could set the side equal to x and then we would at once have a situation where literal symbols must be multiplied, for the area of that square would be x multiplied by x, or xx. Naturally, xx would be the square of x, so although it might seem natural to read xx as "eks eks," it is more common to read it as "eks square."

A similar situation arises in connection with a cube, which is a solid figure with all its angles right angles and all its edges of equal size. Dice and children's blocks are examples of cubes. To obtain the volume of a cube you multiply its length by its width by its height. Since all the edges are of equal size, length, width, and height are all equal. A cube with an edge equal to 2 inches has a volume of $(2)(2)(2)$, or 8 cubic inches.* If it has an edge

No. 8 is on the other side

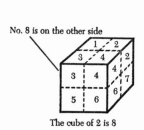

The cube of 2 is 8

Numbers 20 to 26 are on the other side; number 27 is in the center

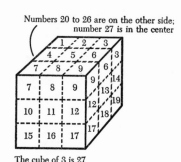

The cube of 3 is 27

---

* I'm sure cubic inches hold no terrors for you. Still, if you feel the need for a little freshening on units of volume, there is always *Realm of Measure* to look through.

equal to 5 inches, it has a volume of (5)(5)(5), or 125 cubic inches.

Because of this, 8 is called the cube of 2, while 125 is the cube of 5. Three equal numbers of any sort multiplied together yield a product that is the cube of the original number.

Again, if we don't know the length of the edge of a cube, and set that length equal to x, we know that the volume is equal to xxx. Naturally, xxx is referred to as "eks cube."

The notions of squares and cubes of a number originated with the Greeks, who were particularly interested in geometrical figures. There are no geometrical figures, however, that can be drawn or built to represent situations where four or more equal numbers must be multiplied, so there are no special names for $xxxx$ or $xxxxx$. Mathematicians simply refer to such expressions as "x to the fourth," "x to the fifth," and so on. If I said "x to the seventeenth" you would know promptly that I meant seventeen x's multiplied together.

In the early days of algebra, mathematicians found they had to get involved frequently with a number of x's multiplied together, and they naturally looked about for some simple and convenient way of symbolizing such a situation. To write out a series of x's takes up space and can be confusing. If you look at $xxxxxxx$, you can't tell at a glance whether you are dealing with x to the seventh, x to the eighth, or x to the ninth. You would have to stop and count them.

A convenient shorthand for this sort of thing was invented by the French mathematician René Descartes (re-NAY day-KART), in 1637. He showed the number of x's to be multiplied together by using a little number placed to the upper right of the x. For instance, xx would be written x^2 and xxx would be written x^3. These are still read "x square" and "x cube." After that, you would have x^4, x^5, or even x^{216}, for that matter, and these are still read "x to the fourth," "x to the fifth," or "x to the two hundred and sixteenth." The little number in such an expression is called the "exponent."

HANDLING THE INVOLUTION

When we consider an expression like x^3 or x^5, we say we are "raising x to a power." This is another algebraic operation.

So far, we have considered addition, subtraction, multiplication, and division — the four algebraic operations that are commonly used in ordinary arithmetic. Raising to a power is a fifth operation, and one that is not commonly used in ordinary arithmetic. This fifth operation can also be called "involution."

You may wonder if it's fair to call involution a

fifth operation. Is it not only a multiplication? Isn't x^2 just x multiplied by x?

So it is, but by the same token, multiplication can be looked upon as a kind of addition, can't it? For instance, $2x$, which is an example of multiplication, can be looked on as simply x plus x, which is certainly an addition.

One reason multiplication is considered a separate operation is that it is handled differently in equations as compared with addition. If we consider the equation $2x + 3 = 10$, we know that we must transpose the 3 before we transpose the 2; that we cannot deal with the multiplication until after we have dealt with the addition.

We could write the equation $x + x + 3 = 10$ and it would be the same equation, but now it would involve only addition. You can now transpose any term at will; you can transpose one of the x's, separating it from the other. Thus, you could change the equation to read $x + 3 = 10 - x$. (This wouldn't do you any good as far as solving for x is concerned, but at least it would leave the equation consistent with what it was before.)

It also turns out, then, that involutions must not be handled in the same way that multiplications are, even though involution can be considered a kind of multiplication. That is what makes involution a separate operation. To show what I mean, let's consider a simple expression involving numerical symbols only.

What is the value of $(2)(2)(5)$? The answer, you can see at once, is 20. It doesn't matter whether you first evaluate $(2)(2)$ to get 4 and then evaluate $(4)(5)$ to get 20; or do it in another order, $(2)(5)$ giving you 10 and $(10)(2)$ giving you 20. The answer is the same no matter in what order you multiply.

Suppose, though, you had written not $(2)(2)(5)$ but $(2^2)(5)$. It is the same expression, but now it involves an involution. The value of 2^2 is 4, of course, and if that is multiplied by 5, the answer is 20. But can you multiply 2 by 5 first and then perform the involution? Suppose, just to see what would happen, you try to do this. Well, $(2)(5)$ is 10 and 10^2 is 100, and now you have an inconsistency.

To avoid the inconsistency, involutions must be performed first, before multiplications are performed; just as multiplications are performed before additions.

As an example, let's look at the expression $3x^2$ and ask ourselves what it means. Does it mean $(3x)(3x)$ or $(3)(x)(x)$?

To answer that question, let's remember that I have explained that since multiplications must be performed before additions, a multiplication is always treated as though it were in parentheses.

The expression $3x + 2$ is always treated as though it were $(3x) + 2$. In the same way, since involution must be performed before multiplication, in an expression involving both, it is the involution that is treated as though it were within parentheses.

Therefore, $3x^2$ is really $3(x^2)$, and its meaning is $(3)(x)(x)$. No confusion is possible. It is only because people have gotten into the habit of leaving out the parentheses in expressions like $3x^2$ that the question arises in the minds of beginners.

(By now you may be a little impatient with the way in which parentheses are left out. Why shouldn't they be put in everywhere possible in order to avoid confusion? Well, if they were, equations would be simply cluttered with parentheses. And as you yourself got expert in manipulating equations, you would get tired of them, and start leaving them out yourself.)

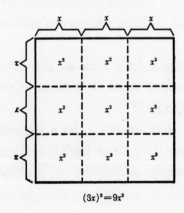

$$(3x)^2 = 9x^2$$

Suppose, though, you really do want to multiply $3x$ by $3x$. How would you indicate that? Again, the very handy parentheses can be used and the multiplication can be written $(3x)^2$. It is easy to see now that it is the entire expression within the parenthesis that is being squared, just as we earlier decided that in the expression $3(x + 2)$, it is the entire portion within the parenthesis that is being multiplied. By proper use of parentheses, we can change the order in which operations must be performed.

If we write out the expression $(3x)^2$ as a multiplication, it becomes $(3)(x)(3)(x)$. Here only multiplications are involved and we can arrange the symbols any way we choose without altering the over-all value. We can therefore write the expression as $(3)(3)(x)(x)$, and this we can change back into involutional form as $(3^2)(x^2)$. If you'll try the same trick on $(7x)^5$, you will find that can be written as $(7^5)(x^5)$.

In fact, we can write that

$$(ab)^n \equiv (a^n)(b^n)$$

where n is the usual general symbol for an exponent.

Using this same line of argument, you should have no trouble in seeing that $(5x)(3x)$ is equal to $15x^2$ and that $30x^3$ can be factored as $(2x)(3x)(5x)$, or even as $(2)(3)(5)(x)(x)(x)$.

COMBINING INVOLUTIONS

What if we are presented with an operation involving two involutions. As an example, consider the expression $(x^3)(x^2)$, where one expression containing an involution is multiplied by another. We can simplify this by substituting multiplication for involution. Thus, x^3 is xxx and x^2 is xx. Therefore, $(x^3)(x^2)$ is $(xxx)(xx)$. Since only multiplications are involved, there is no need for parentheses and the expression becomes $xxxxx$ or x^5. To put it briefly, $(x^3)(x^2) = x^5$. Apparently, we have just added the exponents.

Try other examples and you will find this continues to be so. Exponents are added in all cases of this kind. In general, we can write:

$$(a^m)(a^n) \equiv a^{(m+n)}$$

The obvious next step is to divide one expression involving an involution by another. Can you calculate the value of $\dfrac{x^5}{x^3}$?

You might quickly suppose that since the multiplication of involutions involves addition of exponents, the division of involutions ought to involve the subtraction of exponents. Thus to divide x^5 by x^3 would give the answer x^2, since $5 - 3$ is equal to 2.

The expression x^5 can be written $xxxxx$ and x^3 can

be written xxx. Therefore, $\dfrac{x^5}{x^3}$ is $\dfrac{xxxxx}{xxx}$. We can
cancel three of the x's above and below since each
x represents a common factor. In this way the
three x's in the denominator are subtracted from
the five x's in the numerator and two are left, so
that the answer is indeed x^2.

In general, then:

$$a^m/a^n \equiv a^{(m-n)}$$

This state of affairs can lead us to some interesting conclusions.

Descartes used exponents only when two or more
identical symbols were multiplied together. The
smallest exponent is ^{2} under those conditions. But
suppose you wanted to divide x^3 by x^2? According
to the rules of subtraction of exponents, $\dfrac{x^3}{x^2}$ equals
x^1, but what does that mean?

Well, if $\dfrac{x^3}{x^2}$ is written $\left(\dfrac{xxx}{xx}\right)$, then two of the x's
can be canceled above and below and the value of
the fraction is simply x. Therefore, $\dfrac{x^3}{x^2} = x$. So we
have two answers to the same problem, x^1 and x,
depending on which route of solution we use. The
way to keep consistent is to decide that the two
answers are really one answer by saying that $x^1 \equiv x$.

In a way, this makes sense because, by Descartes' rule, x^1 should represent a single x multiplied together. Even though you can't really multiply an x all by itself (it takes two to multiply), you can imagine it as just standing there untouched so that x^1 should equal x. All this would hold true for any symbol and so we can say

$$a^1 \equiv a$$

But we can go further. Suppose you were to divide x^3 by x^3? Since any number divided by itself equals 1, we can say that $\frac{x^3}{x^3} = 1$. However, if we use our subtraction-of-exponents rule, we must also say that $\frac{x^3}{x^3} = x^{(3-3)}$ or x^0.

Again, the only way to avoid an inconsistency is to agree that the two results are the same and that x^0 is equal to 1. We can work out the same result if we divided 2^3 by 2^3, or 7^5 by 7^5. We must come to the conclusion that any symbol, whether numerical or literal, raised to the zeroth power, equals one. We can say, generally:

$$a^0 \equiv 1$$

Here is an example which shows that consistency in mathematics is more important than "common sense." You might think that x^0 should represent zero x's multiplied together. That doesn't really make sense, but to most people that would *sound* as though the answer should be zero.

But if we let x^0 equal zero, we have the inconsistency that $\dfrac{x^3}{x^3}$ can equal either 1 or 0, depending on which rule of division we use. We simply cannot allow that, so we make x^0 equal 1, regardless of any "common sense" that tells us differently.

And we can go further than that, too. Suppose we want to divide x^3 by x^5. Using the subtraction-of-exponents rule, we have $\dfrac{x^3}{x^5}$ is equal to $x^{(3-5)}$ or x^{-2}. Now what in the world does a negative exponent mean? How can you possibly multiply -2 x's together?

But suppose we write out the expression $\dfrac{x^3}{x^5}$ in a fashion that involves only multiplications, thus, $\dfrac{xxx}{xxxxx}$. Now if we cancel three of the x's, top and bottom, we are left with $\dfrac{1}{xx}$ or $\dfrac{1}{x^2}$.

Again, we have two values for the same expression, x^{-2} and $\dfrac{1}{x^2}$. We must, therefore, set one equal to the other. If we try this sort of thing with other exponents, we would find that x^{-7} must be set equal to $\dfrac{1}{x^7}$; x^{-11} must be set equal to $\dfrac{1}{x^{11}}$, and so on.

In general:

$$a^{-n} \equiv \frac{1}{a^n}$$

Having taken care of the multiplication and division of expressions involving involutions, how about involuting an involution? That sounds odd, but I can explain what I mean in a moment. Suppose you wanted to multiply x^3 by itself. The result would be $(x^3)(x^3)$ or $(x^3)^2$, which is an example of the involution of an involution.

But by the addition-of-exponents rule, $(x^3)(x^3)$ is equal to x^6. Therefore, $(x^3)^2 = x^6$. In the same way $(x^4)^3$ is $(x^4)(x^4)(x^4)$ or x^{12}. In the involution of involutions, we seem to be multiplying exponents and we can say the general rule is

$$(a^m)^n \equiv a^{mn}$$

Thus, the use of exponents simplifies operations. Involution of involution becomes multiplication of exponents. Multiplication of involutions becomes addition of exponents. Division of involutions becomes subtraction of exponents.*

---

* The use of exponents in this fashion led to the invention of "logarithms" as a way of simplifying calculations. This was achieved about 1600 by a Scottish

THE INVERSE OF INVOLUTION

The algebraic operations, other than involution, exist in pairs. Addition has its inverse in subtraction and multiplication has its inverse in division. It wouldn't seem right to have involution exist by itself and without an inverse.

What then is the inverse of involution? Well, let's see. To multiply a quantity by itself thus, (3)(3), is involution, and the product is 9. To construct an inverse operation, we need only begin with the product and work backward. What number multiplied by itself is 9? The answer, of course, is 3.

Or we might ask: What number, taken five times and multiplied together, will give 1024. We can try to answer this hit-and-miss. We might try 2 first, but (2)(2)(2)(2)(2) comes to only 32. In the same way, five 3's multiplied together would give us 243. However, five 4's multiplied together would indeed give us 1024, so our answer is 4.

(There are methods for working out such problems in better ways than hit-or-miss. In this book, I won't deal with that. For the purpose of explaining algebra, I need only simple problems of this

mathematician named John Napier. I have no room to talk about logarithms here, but if you are curious you will find them explained in Chapter 8 of *Realm of Numbers*.

kind — problems so simple you will be able to work them out in your head.)

Mathematicians refer to this inverse of involution, this finding of a value which when multiplied by itself a number of times gives a known answer, as "evolution." I shall try to use this short term whenever possible, but an older and much better-known name for the operation is "extracting a root."

This old-fashioned term comes from the Arabic mathematicians of the Middle Ages. I suppose they looked upon the number 1024 as growing out of a series of 4's as a tree grows out of its roots. Therefore, 4 is a root of 1024. Then, just as tree roots must be extracted from the soil, so 4 must be "extracted" from 1024.

Of course, there are different degrees of roots. Since $(4)(4)$ is 16, $(4)(4)(4)$ is 64, and $(4)(4)(4)(4)$ is 256, 4 is a root of 16, 64, and 256, as well as of 1024. These different situations are distinguished in the same way that different powers are distinguished in involution.

Thus, since 16 is the square of 4, 4 is the square root of 16. Again, 64 is the cube of 4, so 4 is the cube root of 64; 256 is the fourth power of 4, so 4 is the fourth root of 256, and so on.

The operation of evolution is indicated by means of a sign called a "radical" (which has nothing to do with politics but simply comes from a Latin word, "radix," meaning "root"). The sign looks like this: $\sqrt{}$. It was invented by a German mathematician named Christoff Rudolff who used it first in a book published in 1525. Before that, the letter r (for "radix") was used and it is quite possible that $\sqrt{}$ is just a kind of distorted r.

By 1700, mathematicians came to distinguish one kind of root from another by using a little number, just as in the case of involution. Thus, the cube root is written $\sqrt[3]{}$; the fourth root is written $\sqrt[4]{}$; the eighth root $\sqrt[8]{}$; and so on. The little number is referred to as the "index."

By using this radical sign, we can show how powers and roots are related in a very simple way:

$$\text{If} \qquad a^n = b$$

$$\text{then} \qquad a = \sqrt[n]{b}$$

There is one exception to this general rule of indexes, and that involves the square root. It should be written $\sqrt[2]{}$, if we were to be completely logical. However, the square root is used so much oftener than all other kinds of roots put together that mathematicians save time by taking this particular index for granted, just as they usually leave out the 1 in expressions such as $1x$, $\dfrac{x}{1}$, and x^1.

In other words, the sign $\sqrt{}$, standing by itself and without an index, is assumed to be the square root. This is so common, in fact, that the sign is almost never called the radical. It is almost always called the "square-root sign."

Inverse operations always introduce new difficulties. It is subtraction, not addition, that introduces negative numbers. It is division, not multiplication, that introduces fractions.

What new complications will evolution introduce?

Suppose we take the apparently simple problem of finding the square root of 2, or, to use symbols, $\sqrt{2}$. The answer isn't 1 because (1)(1) gives a product of 1, which is less than the desired quantity, 2. The answer isn't 2 either because (2)(2) is 4, which is more than the desired quantity. The answer, then, must lie somewhere between 1 and 2, and if you wish you can try various fractions in that range.

For instance, $\frac{7}{5}$ is almost right since $\left(\frac{7}{5}\right)\left(\frac{7}{5}\right)$ gives the produce $\frac{49}{25}$ or 1.96. It is necessary to find a fraction then that is just a trifle greater than $\frac{7}{5}$.

Unfortunately, if you were to keep on trying, you would never find the correct fraction. Every one you tried would end up just a little higher than 2 if multiplied by itself, or a little lower than 2.

It would never come out exactly 2. Thus, the fraction $\frac{707,107}{500,000}$ if multiplied by itself would give a product of 2.000001237796, which is just a hair above 2, but isn't 2 exactly.

It was the Greeks who first discovered that there were numbers, such as the $\sqrt{2}$, which could not be expressed as fractions, and they were quite disturbed about it. Such a number is now called an "irrational number." Of course, in ordinary speech, "irrational" means crazy or mentally unbalanced and perhaps you think this is a good name for such numbers. In mathematics, however, the name merely means "without a ratio," ratio, you may remember, being another word for fraction.*

Almost all roots, with very few exceptions, are irrational. In this book, I will be constantly using those few exceptions as material to work with in order to keep out of complications. However, don't let that give you the wrong idea. Roots and irrational numbers go hand in hand just as division and fractions do.

---

* I could very easily go on talking about irrational numbers for many pages, but I won't. If you are curious to know more about them, you will find a discussion in *Realm of Numbers*.

EXPONENTS GO FRACTIONAL

In previous cases, we have always managed to get rid of inverse operations in one way or another. By using negative numbers, we got rid of subtractions, for instance, writing $8 + (-7)$ instead of $8 - 7$. Again, we got rid of division by using reciprocals, writing $(4) \left(\dfrac{1}{2}\right)$ instead of $4 \div 2$.

It would seem that we ought to be able to get rid of evolution as well.

To do that, let's begin by considering an expression such as $x^{\frac{1}{2}}$, where a fractional exponent is involved. Do not feel disturbed at this or begin to wonder how half an x can be multiplied together. Remember that the notion of having exponents tell us how many numbers are to be multiplied together is too narrow. We've already gone beyond that in considering expressions like x^0 and x^{-2} and made sense of those impressions. Why not find out how to make sense of $x^{\frac{1}{2}}$ as well?

To begin with, let's multiply $x^{\frac{1}{2}}$ by itself. By the rule of addition of exponents, $(x^{\frac{1}{2}})(x^{\frac{1}{2}})$ is equal to $x^{(\frac{1}{2}+\frac{1}{2})}$ which makes it x^1, or simply x.

If we were to ask then: What number multiplied by itself gives us x? we would have to answer $x^{\frac{1}{2}}$. But in asking what number multiplied by itself gives us x, we are asking: What is the square root of x? So we must say that the square root of x is $x^{\frac{1}{2}}$, or that $\sqrt{x} = x^{\frac{1}{2}}$.

That gives us the meaning of the fractional exponent $\frac{1}{2}$. It is another way of symbolizing the square root.

In the same way, we would find that $(x^{\frac{1}{3}})(x^{\frac{1}{3}})(x^{\frac{1}{3}})$ is equal to x, so that $x^{\frac{1}{3}}$ is the cube root of x, or $\sqrt[3]{x} = x^{\frac{1}{3}}$.

We can make this general by saying

$$\sqrt[n]{a} \equiv a^{\frac{1}{n}}$$

Now you have all you need to understand what is meant by an expression like $x^{2\frac{1}{2}}$. This can be written as $x^{\frac{5}{2}}$, and we know that this is the same as $(x^5)^{\frac{1}{2}}$, because by the rule of multiplication of exponents, $(x^5)^{\frac{1}{2}}$ is equal to $x^{(5)(\frac{1}{2})}$ or $x^{\frac{5}{2}}$.

Since the exponent $\frac{1}{2}$ indicates the square root, we can write $(x^5)^{\frac{1}{2}}$ as $\sqrt{x^5}$. In other words, $x^{\frac{5}{2}}$ equals $\sqrt{x^5}$, and, in general,

$$a^{\frac{m}{n}} \equiv \sqrt[n]{a^m}$$

This way of shifting back and forth from indexes to exponents can temporarily eliminate evolution and make the multiplication of roots simpler. You might well be puzzled, for instance, at being asked to multiply the square root of x by the cube root of x, if all you could do was write it thus: $\sqrt{x} \ \sqrt[3]{x}$.

Change the indexes to exponents, however, and you have $x^{\frac{1}{2}}x^{\frac{1}{3}}$ instead and by the rule of addition of exponents you have the answer $x^{(\frac{1}{2}+\frac{1}{3})}$ or $x^{\frac{5}{6}}$, which can be written as $\sqrt[6]{x^5}$.

And now you can heave a sigh of relief. There will be no new surprises sprung upon you in the way of algebraic operations, for there are no more. Only three pairs of operations exist in the whole of algebra:

      (1) Addition and subtraction

      (2) Multiplication and division

      (3) Involution and evolution

and you now have them all.

The next step is the matter of handling equations that involve involution and evolution.

9

Equations by Degrees

SUPPOSE THAT we had a cube with a volume of 27 cubic inches and were anxious to know the length of the edge. We consider the edge of the cube to be x inches long and, since the volume of a cube is obtained by raising the length of the edge to the third power, we have the equation

$$x^3 = 27$$

In earlier chapters, we found that the same number could be added to or subtracted from both sides of an equation; and that the same number could be used to multiply both sides or to divide into both sides. It isn't hard to suppose that involution and evolution can be added to the list. Both sides of the equation can be raised to the same power or reduced to the same root without spoiling the equation.

In order to change x^3 to x, we need only take the cube root of x^3. The cube root of x^3 is $\sqrt[3]{x^3}$, which is the same as $x^{\frac{3}{3}}$, which, of course, equals x^1 or, simply x.

But if we take the cube root of one side of the equation, we must take the cube root of the other, too, to keep it an equation. Therefore:

$$\sqrt[3]{x^3} = \sqrt[3]{27}$$

The left-hand side of the equation is equal to x, as we have just decided, and the right-hand side is equal to 3, since $(3)(3)(3)$ is equal to 27. The equation becomes

$$x = 3$$

which is the solution.

Suppose, on the other hand, we have an equation like this:

$$\sqrt[3]{x} = 4$$

Now in order to convert $\sqrt[3]{x}$ to x, we cube $\sqrt[3]{x}$. After all, the cube of the cube root of x is $(\sqrt[3]{x})^3$, which can again be written as $x^{\frac{3}{3}}$, or simply x.

In fact, we can set up the general rule for any power or index by this line of argument:

$$\sqrt[n]{a^n} \equiv (\sqrt[n]{a})^n \equiv a$$

If we return to our equation, $\sqrt[3]{x} = 4$, and cube both sides, we have

$$(\sqrt[3]{x})^3 = 4^3$$

or

$$x = 64$$

Just as in the case of the other operations, we are solving equations involving involution and evolution by shifting a symbol from one side of the equation to the other. We are shifting the little figure that represents the index of an evolution or the exponent of an involution. If you look at the equations so far in this chapter again, you will see that

$$\text{If} \qquad x^3 = 27$$

$$\text{then} \qquad x = \sqrt[3]{27}$$

$$\text{And if} \qquad \sqrt[3]{x} = 4$$

$$\text{then} \qquad x = 4^3$$

In each case, you might say the little ³ has been transposed. (Actually, the term "transpose" is confined to the operations of addition and subtraction. However, I have used it for multiplication and division as well and the shift in the case of evolution and involution is so similar in some ways, that I will even use the word here.) In this case, too, as with the other operations, transposition means an inversion. Involution becomes evolution as the exponent is converted to an index. And evolution becomes involution as the index is converted to an exponent. This is plain enough in the samples I have just given you.

The reason this isn't as plain as it should be is that almost all the roots used in algebra are square roots, and this is the one case where the index is omitted.

If we say, for instance, that

$$x^2 = 16$$

you can see at once by what I have said so far that

$$x = \sqrt{16}$$

but now the little ^{2} seems to have disappeared in the process, and the fact that it has been transposed and changed from exponent to index is not noticeable. It would be noticeable if we wrote the square root of 16 as $\sqrt[2]{16}$, as we should logically do — but which we don't.

In the same way, if we start with

$$\sqrt{x} = 8$$

we can convert that to

$$x = 8^2$$

and a little ^{2} seems to have appeared out of nowhere.

SOLUTIONS IN PLURAL

Mathematicians are very conscious of the appearance of powers in equations. They make equations more difficult to handle. The higher the powers, in fact, the more difficult equations are to handle. By 1600, therefore, mathematicians were classifying equations according to the highest power of the unknown that appeared in them.

Equations are said to be of a certain "degree." In a simple equation of the type I have used up to this chapter, such as $x - 3 = 5$, the unknown can be written x^1, so this is an "equation of the first degree."

In the same way, an equation such as $x^2 - 9 = 25$ is an "equation of the second degree" because the unknown is raised to the second power. In an equation such as $x^2 + 2x - 9 = 18$, where the unknown is raised to the second power in one term and to the first power in another, it is the higher power that counts and the equation is still of the second degree.

Equations such as $x^3 - 19 = 8$, or $2x^3 + 4x^2 - x = 72$, are "equations of the third degree" and so on.*

---

* Actually, these should be referred to as "polynomial equations" of this degree or that, because they involve polynomials, whereas some kinds of equations do not. However, in this book, I talk about polynomial equations and no other kind, so I won't bother to specify all the time.

This is the simplest and most logical way of classifying equations, but, unfortunately, mathematicians have become so familiar with these different types that they have also given them specialized names. Since the specialized names are used more often than the simple classification by degree, you had better be told what they are.

An equation of the first degree is called a "linear equation" because the graph of such an equation is a straight line. (Graphs are, alas, not a subject I can cover in this book.)

An equation of the second degree is called a "quadratic equation," from the Latin word "quadrus," meaning "square." A quadratic equation is an equation involving squares, after all, so that's fair enough.

An equation of the third degree, with even more directness for English-speaking people, is called a "cubic equation."

An equation of the fourth degree is called a "quartic equation," and one of the fifth degree is called a "quintic equation," from the Latin words for "four" and "five" respectively. Sometimes an equation of the fourth degree is called "biquadratic," meaning "two squares" because it involves the multiplication of two squares. After all, $(x^2)(x^2) = x^4$.

Having settled that, then, let's take a close

look at the simplest possible quadratic equation:*

$$x^2 = 1$$

By transposing the exponent, we have

$$x = \sqrt{1}$$

and all we need ask ourselves is what number multiplied by itself will give us 1. The answer seems laughably simple since we know that (1)(1) is equal to 1, so

$$x = 1$$

But hold on. We haven't been worrying about signs. When we say that $x^2 = 1$ and that $x = \sqrt{1}$, what we really mean is that $x^2 = +1$, and that $x = \sqrt{+1}$. (The plus sign, remember, is another one of the many symbols that mathematicians keep omitting.)

This changes things. If we ask what number multiplied by itself will give us $+1$, we are suddenly in a quandary. It is true that $(+1)(+1)$ is equal to $+1$, but isn't it also true that $(-1)(-1)$ is equal to $+1$? Therefore, are not $+1$ and -1 both solutions for x in the equation $x^2 = 1$?

Does this sound like an inconsistency to you,

---

* It might seem to you that $x^2 = 1$ is not a polynomial equation because no polynomials are involved. However, by transposing, you have $x^2 - 1 = 0$ and there's your polynomial.

with two answers for the value of x in one equation? You might try to remove this troublesome complication by just deciding on a rule that won't count negative numbers as a solution to an equation. Mathematicians up into the 1500's actually did use such a rule.

However, they were wrong to do it. Negative numbers are so useful that to eliminate them merely to avoid a complication is wrong. Besides, the rule does no good. There are quadratic equations which give two answers that are both positive.

For instance, take the equation $x^2 - 3x = -2$. Without actually going through the procedure of solving for the value of x, I will simply tell you that both 1 and 2 are solutions. If you substitute 1 for x, then x^2 is 1, and $3x$ is 3, and $1 - 3$ is indeed -2. If, however, you substitute 2 for x, then x^2 is 4 and $3x$ is 6, and $4 - 6$ is also equal to -2.

There are other quadratic equations in which two solutions exist that are both negative. In the equation $x^2 + 3x = -2$, the two solutions for x are -1 and -2.

By 1600, mathematicians had resigned themselves to the thought that the rules for quadratic equations weren't the same as those for linear equations. There could be two different solutions for the unknown in a quadratic equation.

Well, then, are we stuck with an inconsistency?

Of course not. An inconsistency results when solving an equation by one method yields one answer, and solving it by another yields a second answer. Solving a quadratic equation gives two answers *at the same time*. And no matter what different methods you use, you get the same two answers.

To make this quite plain, if I were to ask you the name of the largest city in the United States, and you answered New York one time and Chicago another time, you would be inconsistent. If, however, I asked you the name of the two largest cities in the United States and you answered New York and Chicago, you would *not* be inconsistent. You would be correct.

In fact, to return to equations, it was soon discovered that the unknown in a cubic equation could have three solutions and the unknown in a quartic equation could have four solutions. In 1637, René Descartes, the man who invented exponents, decided that the unknown of any equation had a number of solutions exactly equal to the degree of the equation. This was finally proved completely in 1799 by a German mathematician named Carl Friedrich Gauss. (His name is pronounced "gows.")

IMAGINARY NUMBERS

But if this is so, and the unknown in a cubic equation has three solutions, then in the very simplest cubic equation (in which I will include signs this time)

$$x^3 = +1$$

there should be three solutions for x.

Transposing the exponent, we have

$$x = \sqrt[3]{+1}$$

and we need only ask ourselves what number can be taken three times and multiplied to give $+1$. We can start off instantly by saying that $+1$ is itself a solution since $(+1)(+1)(+1)$ is equal to $+1$.

But where are the other two solutions? Can one of them be -1? Well, $(-1)(-1)$ is $+1$ and multiplying that product by a third -1 gives us $(+1)(-1)$, which is equal to -1. Therefore $(-1)(-1)(-1)$ is equal to -1 and not to $+1$, so that -1 is not a solution to the cubic equation above.

Nor is there any number, any fraction, or even any irrational, either positive or negative, with the single exception of $+1$, that is the cube root of $+1$.

Then what conclusion can we come to but that here we have a cubic equation in which the unknown has but a single solution?

Can it be that we have caught great mathematicians such as Descartes and Gauss in an error? And so quickly and easily?

That's a little too good to believe, so let's go back a step to the quadratic equation again. Now that we're introducing signs, let's try an equation of this sort with a negative, like this:

$$x^2 = -1$$

By transposing the exponent, we have

$$x = \sqrt{-1}$$

and now we must find a number which, multiplied by itself, equals -1. Since both $(+1)(+1)$ and $(-1)(-1)$ equal $+1$, we are suddenly left with the thought that there is *no* number which is the square root of minus one. Can it be that in an equation as simple as $x^2 = -1$, we are faced with no solution at all?

In 1545, however, Cardano, who, you may remember, introduced negative numbers, decided to invent a number which, when multiplied by itself gave -1 as the product. Since this number didn't seem to exist in the real world but only in imagination, he called this number an "imaginary number." In 1777, the Swiss mathematician Leonhard Euler (Oi-ler) symbolized the square root of -1 as i (for "imaginary").

In other words, i is defined as a number which, when multiplied by itself, gives -1. You can write this as $(i)(i) = -1$, or $i^2 = -1$, or $i = \sqrt{-1}$.

It seems difficult for people to accept i as being a number that is just as valid as 1. It doesn't help to call i "imaginary" and numbers like $+1$ or -1 "real numbers," but this is what is done to this day.

There is, in actual fact, nothing imaginary about i. It can be dealt with as surely as 1 can be dealt with. Thus, you can have two different kinds of i just as you can have two different kinds of 1. You can have $+i$ and $-i$. Just as $+1$ and -1 have the same square, $+1$, so $+i$ and $-i$ have the same square, -1.

Thus the square root of -1 has two solutions rather than none, the solutions being $+i$ and $-i$. Often, when an unknown is equal to both the positive and negative of a particular number, the number is written with a sign made up of both the positive and negative, thus $\pm$. This is read "plus or minus" so that the number ± 1 is read "plus or minus one." Using this sign:

$$\text{If} \qquad x^2 = -1$$

$$\text{then} \qquad x = \pm i$$

Are you still anxious for some way of visualizing what i is and of getting its exact nature clear in your mind? Well, remember that at first people had this trouble with negative numbers. One way of explaining what the mysterious less-than-zero numbers might be was to use directions. Thus, +1 might be represented by a point 1 inch east of a certain starting position while +2 might be represented by a point 2 inches east of the starting position, +5 by a point 5 inches east, and so on. In that case, negative numbers would be represented by westward positions, and −1 would be represented by a point 1 inch west of the starting position.

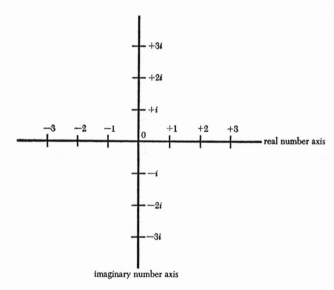

real number axis

imaginary number axis

We can follow right along with this and have +i and the other positive imaginaries be represented by distances to the north and −i and the other negative imaginaries represented by distances to the south.

Imaginaries actually are represented in this way and have helped mathematicians tremendously. With only real numbers, mathematicians were confined to a single east–west line, so to speak. Once imaginary numbers were introduced, they could wander in all directions — east, west, north, and south. It was just like being let out of prison, and what is called "higher mathematics" would be impossible without imaginary numbers.

The symbol **i** is all that is needed in dealing with imaginary numbers. You might think, offhand, that there would be an unending number of symbols required for the unending number of imaginaries. For instance, $\sqrt{-4}$, $\sqrt{-8}$, $\sqrt{-16}$, and so on are all imaginary, for no real number of any sort will give a negative number of any sort as a square.

To see why **i** is nevertheless sufficient, let's look first at $\sqrt{36}$, which equals 6. The expression can be written $\sqrt{(9)(4)}$, and if this is broken apart into the multiplication of two separate roots, thus, $(\sqrt{9})(\sqrt{4})$, we can evaluate it as $(3)(2)$, which is still equal to 6. Other examples will show that this is a general rule and that

$$\sqrt{ab} = (\sqrt{a})(\sqrt{b})$$

Now we can write $\sqrt{-4}$ as $\sqrt{(4)(-1)}$, which, by the rule just cited, can be written as $(\sqrt{4})(\sqrt{-1})$ which works out to $(2)(\mathbf{i})$, or simply $2\mathbf{i}$. In the same way, $\sqrt{-16}$ works out to 4i, and $\sqrt{-8}$ equals $\sqrt{8}$ i. (The value of $\sqrt{8}$ is an irrational number, but a close value is 2.824, so we can say that $\sqrt{-8}$ is about equal to 2.824 i.)

In fact, we can state the general rule that

$$\sqrt{-n} \equiv i\sqrt{n}$$

and for that reason, the existence of i takes care of the square roots of all negative numbers.

We can now say that the unknown of any quadratic equation has two solutions, provided we remember that the solutions need not be real numbers, but might be imaginaries.

We can even venture into a quartic equation such as

$$x^4 = +1$$

By transposing the exponent, we have

$$x = \sqrt[4]{+1}$$

and we need to find values for x such that four of them multiplied together give $+1$. Two possible values for x are $+1$ and -1, since $(+1)(+1)(+1)(+1) = +1$, and $(-1)(-1)(-1)(-1) = +1$. If you're not certain that the latter multiplication is correct, think of it this way: $[(-1)(-1)][(-1)(-1)]$. The first two -1's multiplied together give $+1$ as the product and so do the last two -1's. The expression therefore becomes $(+1)(+1)$, or $+1$.

Two more possible values for x are $+i$ and $-i$. In the first place, consider $(+i)(+i)(+i)(+i)$. Break that up into pairs as $[(+i)(+i)][(+i)(+i)]$ and you see that the product of the first pair is -1, and so is the product of the last pair. The expression therefore becomes $(-1)(-1)$, which gives the desired result of $+1$. By the same reasoning $(-i)(-i)(-i)(-i)$ is also equal to $+1$.

So you see that in the quartic equation $x^4 = +1$, there are indeed four possible solutions for x, these being $+1$, -1, $+i$, and $-i$. Imaginary numbers are thus essential for finding the proper number of solutions. But are they sufficient?

COMBINING THE REAL AND IMAGINARY

Let's go back to the cubic equation

$$x^3 = +1$$

where, so far, we have found only one solution for x, that solution being $+1$ itself. If we allow imaginary numbers, can it be that $+i$ and $-i$ are the second and third solutions?

What about the expression $(+i)(+i)(+i)$? The first two $+i$'s, multiplied together, equal -1, so the expression becomes $(-1)(+i)$ which, by the law of signs, gives the product $-i$. Therefore, $+i$ is not a solution of the equation. In the same

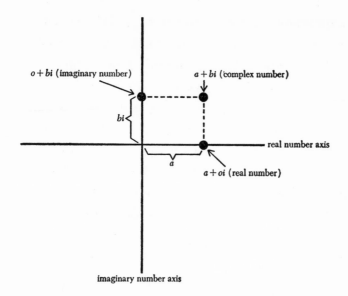

way, you can work out that $(-i)(-i)(-i)$ yields the product $-i$, so $-i$ is not a solution of the equation either. Where, then, are the second and third solutions of the equation $x^3 = +1$?

To find the answer to that question we must realize that most numbers are neither entirely real nor entirely imaginary. They're a combination of both. Using general symbols, we can say that the typical number looks like this: $a + bi$, where a is a real number and bi is an imaginary number.

Such numbers, part real and part imaginary, were called "complex numbers" by Gauss in 1832.

Actually, those numbers that are entirely real or entirely imaginary can also be considered as examples of complex numbers. In the expression $a + b\mathbf{i}$, suppose a is set equal to 0. The expression becomes $0 + b\mathbf{i}$, or just $b\mathbf{i}$. By letting b equal any value except 0, any imaginary number can be produced. All the imaginary numbers, then, are complex numbers of the form $0 + b\mathbf{i}$.

Suppose, on the other hand, b were set equal to 0, in the expression $a + b\mathbf{i}$. The expression becomes $a + 0\mathbf{i}$. But $0\mathbf{i}$ is 0, since any number multiplied by zero is zero, so that the expression can be written $a + 0$, or simply a. By letting a equal any value except 0, any real number can be produced. All the real numbers, then, are complex numbers of the form $a + 0\mathbf{i}$.

The solutions of the equation $x^2 = +1$ can be given as complex numbers. They are $1 + 0\mathbf{i}$ and $-1 + 0\mathbf{i}$. As for $x^2 = -1$, the two solutions, as complex numbers, are $0 + \mathbf{i}$ and $0 - \mathbf{i}$.

But the complex numbers that serve as solutions for x in a particular equation do not have to have zeros involved every time. A solution can consist of a complex number, $a + b\mathbf{i}$, in which neither a nor b is zero. And, in fact, that is what happens in the cubic equation, $x^3 = +1$. The solution which I've been calling $+1$ can be written in complex form as $1 + 0\mathbf{i}$. The other two solutions don't involve a zero. They are $-\frac{1}{2} + \frac{1}{2}\sqrt{3}\,\mathbf{i}$, and $-\frac{1}{2} - \frac{1}{2}\sqrt{3}\,\mathbf{i}$.

If either of these expressions is cubed, the answer is −1.*

It is by using the system of complex numbers that it is possible to show that x in any equation has a number of solutions exactly equal to the degree of the equation.

Now that this is settled, I will go back to quadratic equations and, for a while at least, we can forget about equations of a higher degree than two.

---

* I'm going to ask you to take my word for this because I have no room to go any further into imaginaries in this book. However, in *Realm of Numbers*, where I discuss imaginaries a bit more in detail, you can find the cube of these expressions worked out toward the end of Chapter 9.

10

Factoring the Quadratic

WE CAN make equations of the second degree a little more complicated by adding other operations, as in

$$3x^2 - 8 = 100$$

I have already explained that a multiplication is treated as though it were enclosed in a parenthesis. The same is true of involution, only more so, so that when multiplication is also present, the involution is enclosed in an inner parenthesis. The equation could be written, if all parentheses were included:

$$[3(x^2)] - 8 = 100$$

Naturally, we must transpose first the 8, then the 3, then the 2, working inward through the layers of parentheses. The results are:

$$3x^2 = 100 + 8 \qquad \text{(or 108)}$$

$$x^2 = \frac{108}{3} \qquad \text{(or 36)}$$

$$x = \sqrt{36} \qquad (\text{or} \pm 6)$$

which are the solutions.

Where parentheses are involved in different arrangement, the order of transposition changes accordingly. In the equation

$$(x + 3)^2 - 7 = 42$$

which could be written, in full, as

$$[(x + 3)^2] - 7 = 42$$

the 7 is transposed first, but the other operation of addition is within an inner parenthesis and can't be dealt with until the exponent in the outer parenthesis is taken care of. So the order of transposition is the 7, then the 2, then the 3, thus:

$$(x + 3)^2 = 42 + 7 \qquad (\text{or } 49)$$

$$x + 3 = \sqrt{49} \qquad (\text{either} + 7 \text{ or} - 7)$$

Therefore, either

$$x = 7 - 3 \qquad (\text{or } 4)$$

which is one solution, or

$$x = -7 - 3 \qquad (\text{or} - 10)$$

which is another solution.

So far, quadratic equations seem just like ordinary linear equations with just the additional complication of having three sets of operations to worry about instead of two, and of having two solutions instead of one.

The two solutions are each perfectly valid, of course. Taking the second case as an example, either numerical value determined for x can be substituted in the equation

$$(x + 3)^2 - 7 = 42$$

Substituting 4, gives you:

$$(4 + 3)^2 - 7 = 42$$

$$7^2 - 7 = 42$$

$$49 - 7 = 42$$

which is correct.
Substituting − 10, gives you:

$$(-10 + 3)^2 - 7 = 42$$

$$-7^2 - 7 = 42$$

$$49 - 7 = 42$$

which is again correct.

And yet there is more to the quadratic equation than you have seen so far. A quadratic equation, remember, might also have a term containing an x to the first power, thus:

$$x^2 + 5x = 6$$

Such an equation can arise very naturally out of a problem such as the following: Suppose you had a rectangular object with its width 5 feet greater than its length and with an area of 6 square feet. What is the length and width of the rectangle?

To begin with, let's set the length equal to x. The width, being 5 feet greater than the length, is, naturally, $x + 5$. Since the area of a rectangle is obtained by multiplying the length by the width, $x(x + 5)$ is that area, which is given as 6 square feet.

By the rules concerning the removal of parentheses, which I gave you earlier in the book, we know that $x(x + 5)$ can be written as $(x)(x) + (x)(5)$, which comes out to $x^2 + 5x$, and that gives us our $x^2 + 5x = 6$.

We can get a sort of solution for x by transposing as follows:

$$x^2 = 6 - 5x$$

$$x = \pm\sqrt{6 - 5x}$$

but that gets us nowhere, for we have x equal to an expression which contains an x, and we can't evaluate it. In fact, we have jumped from the frying pan into the fire, for we have exchanged a power for a root, and the roots are harder to handle.

FROM SECOND DEGREE TO FIRST

A possible way out of the dilemma lies in the
process of factoring. Factoring, as you recall, is
a method of breaking up an expression into two
other expressions which, when multiplied, give you
the original expression. Thus, 69 can be factored
into $(23)(3)$, and $5x - 15$ can be factored into
$(5)(x - 3)$. Factoring generally converts a complicated expression into two or more simpler ones and
there is always the good chance that the simpler
ones can be handled where the original complicated
one cannot.

Naturally, before we can figure out how to factor
an expression, we must have some ideas about the
process of multiplication. If we knew the kind of
multiplications that gave rise to a particular type
of expression, we would know better how to break
that expression apart.

I've already given an example of one multiplication that gives rise to a second-degree term,
when I talked about $x(x + 5)$. That was a multiplication that raised no problems since we know
that in such a multiplication, the term outside the
parenthesis is multiplied by each of the terms inside and that the products are all added. But
suppose I had an expression such as $(x + 7)(x + 4)$?
How does one multiply that?

It might seem logical to go one step further. Take each term in one parenthesis and multiply it by each term in the other, and then add all the products. Would that give the correct answer?

To see if it does, let's go back to ordinary numerical symbols and see what that will tell us. Suppose we were trying to multiply 23 by 14. Actually, 23 is 20 + 3 and 14 is 10 + 4, so (23)(14) can also be written (20 + 3)(10 + 4).

Now let's try multiplication. First, we multiply 20 by each term in the other parenthesis; (20)(10) is 200, and (20)(4) is 80. Doing the same next for the 3, (3)(10) is 30, and (3)(4) is 12. If we add all four "partial products" we have 200 + 80 + 30 + 12, or 322. Multiply 23 by 14 your own way and see if that isn't the answer you get.

Of course, you may say that this isn't the way *you* multiply, but actually it is. You have been taught a quick mechanical rule of multiplying numbers, but if you study it carefully you will find that what you are really doing is exactly what I have just done.

We can write the multiplication of 23 and 14 this way:

with the arrows pointing out all the different multiplications involved. In fact, some people think that the crossed arrows in the center are what gave rise to the sign $\times$ for multiplication.

Now this same system works when literal symbols replace numerical symbols. In the expression $(x + 7)(x + 4)$, we can set up the multiplication this way:

$$x + 7$$
$$x + 4$$

The four multiplications are $(x)(x) = x^2$; $(x)(4) = 4x$; $(7)(x) = 7x$, and $(7)(4) = 28$.

Now how do we add these four partial products? There is no use trying to add a term containing x^2 with one containing x or with one containing no literal symbol.

The x^2 must remain x^2 and, by the same reasoning, the 28 must remain 28. However, the remaining two submultiples both contain x. They are $4x$ and $7x$ and they, at least, can be added to give $11x$. So we end with the following:

$$(x + 7)(x + 4) \equiv x^2 + 11x + 28$$

This means that if you were to come across the expression $x^2 + 11x + 28$, you could at once replace it by $(x + 7)(x + 4)$. The quadratic ex-

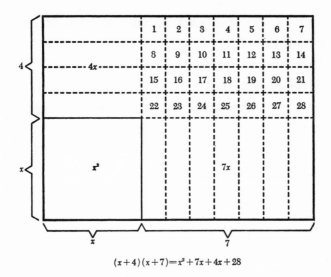

$$(x+4)(x+7)=x^2+7x+4x+28$$

pression would be factored, in this way, into two
expressions, each of which was, by itself, only of
the first degree.

Ah, yes, you may be thinking, but how does one
go about doing that? How can you tell just by
looking at $x^2 + 11x + 28$ that it can be factored
into $(x + 7)(x + 4)$. The only reason we know
about it here is that we did the multiplication first,
and that's like looking at the answers in the back
of the book.

Well, you're right. Factoring is a tricky job and
sometimes it's a hard job. That's true even in
arithmetic. You know that 63 can be factored as
$(7)(9)$, but how do you know that? Only because
you've multiplied 7 by 9 so many times in your life

that $63 = (7)(9)$ is part of your general stock of knowledge.

But can you factor the number 24,577,761?

Most people couldn't just by looking at it, any more than they could factor a polynomial algebraic expression. However, there are some rules to factoring. For instance, the digits in 24,577,761 add up to 39, which is divisible by 3. This means that 3 is one factor of 24,577,761 and it can therefore be factored as (3)(8,192,587).

There are rules that can guide you in factoring algebraic expressions as well.

For instance, suppose we multiply two expressions, using general symbols instead of numerical symbols. We will multiply $x + a$ by $x + b$. The multiplication would look like this:

$$x + a$$
$$x + b$$

The four partial products are x^2, ax, bx, and ab. The ax and bx can be combined as $(a + b)x$, so that we can state the following:

$$(x + a)(x + b) \equiv x^2 + (a + b)x + ab$$

This gives us a pattern. The coefficient of the x term is $a + b$, while the term without the x is ab. We can combine expressions now without even

going to the trouble of multiplying them out. If
we have the multiplication $(x + 17)(x + 5)$, then
we see at once that the coefficient of the x term must
be $17 + 5$, or 22, while the term without the x is
$(17)(5)$ or 85. The product is therefore $x^2 + 22x$
$+ 85$.

To factor a polynomial, we need only work this
backward; look for two numbers which by addition
will give the coefficient of the x term, and by multiplication the numerical term.

Consider the polynomial $x^2 + 7x + 12$. If you
consider 7 and 12, it might occur to you that the
key numbers are 3 and 4, for $3 + 4$ is 7, while
$(3)(4)$ is 12. You can factor the polynomial to
$(x + 3)(x + 4)$. Multiply those factors and see if
you don't get the polynomial.

Or suppose you have the expression $x^2 - 2x$
$- 15$. If you think a while, it may occur to you
that $-5 + 3$ is equal to -2, while $(-5)(3)$ is
-15. These, -5 and 3, are the key numbers, then,
and you can factor the polynomial to $(x - 5)$ and
$(x + 3)$.

Of course, for this to work out properly, the first
term must be x^2 and not $3x^2$ or $5x^2$ or something
like that. Where x^2 has a coefficient other than 1,
factoring can still be performed but it becomes a
trifle more complicated.

Sometimes you can factor an expression that

lacks an x term — one such as $x^2 - 16$, for instance.

To see how that is done, let's multiply general symbols again, but with a small difference. This time we will multiply $x + a$ by $x - a$. We set it up, thus, substituting $x + (-a)$ for $x - a$:

$$x + a$$
$$\downarrow \times \downarrow$$
$$x + (-a)$$

The four partial products are x^2, ax, $-ax$, and $-a^2$. The presence of the $-a$, you see, makes two of the submultiples negative. Now if you add the submultiples together, you get $x^2 + ax - ax - a^2$. The ax and $-ax$ yield zero on addition, so that you end with

$$(x + a)(x - a) \equiv x^2 - a^2$$

Since that is so, an expression such as $x^2 - 16$ is easy to factor. It can be written $x^2 - 4^2$, since 4^2 is equal to 16, and from the pattern of the general equation I have just given you, you see that $x^2 - 4^2$ must factor as $(x + 4)(x - 4)$. Multiply those factors and see if you don't get $x^2 - 16$.

This by no means concludes the rules of factoring. For instance, can you imagine what one must do if one has a coefficient to the x^2 term? How does one factor an expression like $2x^2 + 13x + 15$? That, however, I will leave to you. I have gone as far into factoring as I need to for the purposes of this book.

THE USES OF FACTORING

Of course, you might be asking yourself how you are helped by factoring. What does it do for you as far as handling equations is concerned?

The best way to answer that is to give an example of how factoring will change a seemingly complicated equation into a simple one. Let's begin with the following equation:

$$\frac{x^2 + x - 20}{x^2 - 25} = 15$$

One way to handle it would be to transpose the denominator

$$x^2 + x - 20 = 15(x^2 - 25)$$

remove the parentheses

$$x^2 + x - 20 = 15x^2 - 375$$

bring all the literal symbols to the left and all numerical symbols to the right by appropriate transposition

$$x^2 - 15x^2 + x = -375 + 20$$

or

$$-14x^2 + x = -355$$

We can remove the negative sign at the beginning by multiplying each side of the equation by -1 to get

$$14x^2 - x = 355$$

We might even try factoring the left-hand side of the equation so as to have it read

$$x (14x - 1) = 355$$

but where do we go from there? We are stuck, unless we try different values for x, hit-and-miss, and see which one will solve the equation.

Now let's return to the original equation

$$\frac{x^2 + x - 20}{x^2 - 25} = 15$$

and try factoring before we do anything else. Consider the numerator of the fraction, $x^2 + x - 20$. The coefficient of the x term is $+1$ (a coefficient of 1 is always omitted, of course, but don't forget it's really there just the same and that I warned you earlier in the book you would have to keep it in mind) and the numerical term is -20. Now it so happens that when $+5$ and -4 are added, the sum is $+1$, while if they are multiplied, the product is -20. Therefore the numerator can be factored as $(x + 5)(x - 4)$.

As for the denominator, $x^2 - 25$, that can be written as $x^2 - 5^2$ and can therefore be factored as $(x + 5)(x - 5)$. Now we can write the equation as

$$\frac{(x + 5)(x - 4)}{(x + 5)(x - 5)} = 15$$

But the factor $x + 5$ appears in both the numerator and the denominator and can therefore be canceled.

The equation becomes

$$\frac{x - 4}{x - 5} = 15$$

and suddenly everything is quite simple. We transpose the denominator and proceed according to the usual rules:

$$x - 4 = 15(x - 5)$$

$$x - 4 = 15x - 75$$

$$x - 15x = -75 + 4$$

$$-14x = -71$$

$$14x = 71$$

$$x = \frac{71}{14}$$

which is the solution.

Factoring can help in another way, too. Suppose you have the following equation:

$$x^2 - 9x + 18 = 0$$

Since -3 and -6 will give -9 when added and $+18$ when multiplied, we can factor the expression and write the equation this way:

$$(x - 3)(x - 6) = 0$$

We can now take advantage of an arithmetical fact. Whenever we multiply two factors and find an answer of 0, then one factor or the other must be equal to 0. If neither factor is 0, then the product can never be zero.

Suppose, then, that $x - 3$ is equal to 0. By transposition, we can see at once that if

$$x - 3 = 0$$

then

$$x = 3$$

But it's possible that it is the other factor that is 0; in other words, that

$$x - 6 = 0$$

and

$$x = 6$$

Which, then, is the correct answer? Is x equal to 3 or to 6? There is no reason to suppose that one factor has more chance to be 0 than the other factor, so both answers are equally right. This shouldn't bother us, since the equation is one of the second degree and should have two solutions anyway. We can check (just to make sure) by trying both solutions in the original equation, $x^2 - 9x + 18 = 0$.

First we try the 3, so that the equation becomes $3^2 - (9)(3) + 18$, or $9 - 27 + 18$, which does indeed come out to be 0, so that 3 is a proper solution.

Next the 6 is substituted for x and the equation becomes $6^2 - (9)(6) + 18$, which is $36 - 54 + 18$, which also comes out to 0, so that 6 is another proper solution.

Naturally, we can only take advantage of this device when we can set the product of two factors equal to zero, and to do that, we must begin by getting a zero on the right-hand side of the equation.

For instance, earlier in the chapter I spoke of a rectangular object with a width 5 feet greater than its length and with an area of 6 square feet. The equation involved was:

$$x^2 + 5x = 6$$

At the time, we could go no further with the equation, but now suppose we transpose the 6 to the left in order to leave the very desirable zero on the right. The equation would read:

$$x^2 + 5x - 6 = 0$$

Since the two numbers 6 and -1 give a sum of 5 when added and a product of -6 when multiplied, the expression on the left can be factored and the equation written

$$(x + 6)(x - 1) = 0$$

and our two solutions for x are -6 and 1, these being obtained by setting each factor equal to zero.

The meaning of the solution 1 is clear. The rectangular object has a length of 1 foot. Its width, which is 5 feet greater than its length, is therefore 6 feet, and its area is $(1)(6)$ or 6 square feet, as stated in the problem.

But what about the solution -6? Can we say the length is -6 feet, and the width, which is 5 feet greater, is -1 feet? The area would be (-6) (-1) or still 6 square feet, but what is the meaning of a negative length? The Greeks threw out negative solutions to such equations, feeling certain there was no meaning to a negative length.

However, we can arrange a meaning by supposing measurements made in one direction to be positive and in the other direction, negative. In that case, the two solutions actually apply to the same object placed in two different fashions.

11

Solving the General

THERE IS something unsatisfactory about solving quadratic equations by factoring. After all, some equations cannot be factored very easily. Isn't there a way of solving *any* quadratic equation, without worrying about factors?

To show you what I mean, let's not consider specific equations, but general equations. For instance, here is the "general equation of the first degree," one which uses parameters instead of numbers so that it can represent any such equation:

$$ax + b = 0$$

By allowing a and b to take on any particular values, any particular equation can be represented by this expression. If a is set equal to 2 and b to 3, the equation becomes $2x + 3 = 0$. Subtractions are not excluded because of the plus sign in the general equation, for if a is set equal to 2 and b to -3, the equation becomes $2x + (-3) = 0$, or $2x - 3 = 0$.

Even such an equation as $9(x + 4) = 25$ can be expressed in the general form. By removing the parenthesis and transposing, we have:

$$9(x + 4) = 25$$

$$9x + 36 = 25$$

$$9x + 36 - 25 = 0$$

$$9x + 11 = 0$$

Therefore, the equation $9(x + 4) = 25$ can be put into the general form of $ax + b = 0$ with a equal to 9 and b to 11.

To solve the general equation of the first degree is easy:

$$ax + b = 0$$

$$ax = -b$$

$$x = \frac{-b}{a}$$

This means that if you have any equation in the first degree, you need only put it into the general form and you can obtain the value of x at once. You don't have to worry about transposing, simplifying, or factoring any further. All you have to know are the coefficients, the values of a and b.

Thus, in the equation $17x - 8 = 0$, the value of

x is $\frac{8}{17}$, with no further ado. If it had been $17x + 8 = 0$, the answer would be $-\frac{8}{17}$.

As you see, the value of x in the general equation of the first degree can be any imaginable integer or fraction, if the values of a and b are set at the appropriate whole-number values. If you make up a fraction at random, say, $\frac{17,541}{222,222}$, then that is the value of x in the equation $222,222x - 17,541 = 0$.

We can put this in another way by saying that in the general equation $ax + b = 0$, where a and b are any integers, positive or negative, the value of x can be any rational number. It can never be an irrational number, however, for the answer is bound to come out a definite fraction.

You may wonder what happens if the values of a and b are not integers. Suppose they are fractions. If so, those fractions can always be converted into integers. Thus, in the equation

$$\left(\frac{15}{22}\right) x + \frac{3}{5} = 0$$

suppose you multiply both sides of the equation by the product of the denominators of the two fractions; $(22)(5)$ or 110. We then have

$$\left[\left(\frac{15}{22}\right) x + \frac{3}{5}\right] [110] = (0)(110)$$

Removing parentheses in the usual way and remembering that (0)(110) (or zero times any number, in fact) is 0, we have

$$\left[\left(\frac{15}{22}\right)x\right][110] + \left(\frac{3}{5}\right)(110) = 0$$

$$\left(\frac{1650}{22}\right)x + \frac{330}{5} = 0$$

$$75x + 66 = 0$$

Thus, this equation with fractional coefficients is converted to one with integral coefficients and the solution of x is $-\frac{66}{75}$. Any equation in any degree can be converted from fractional coefficients to integral coefficients, so we need only consider the latter case.

Suppose, though, that a and b were not rational numbers at all (that is, not fractions) but were irrationals, as in the equation $\sqrt{2}\,x - \sqrt{3} = 0$. The value of x would then be $\frac{\sqrt{3}}{\sqrt{2}}$ and it would itself be irrational. However, if we consider only general polynomial equations with rational coefficients, then we can say that the value of x in an equation of the first degree can be any rational number, but cannot be an irrational number.

Having gone through all this, which is pretty straightforward, we are ready to ask how we might handle the general equation of the second degree. How could we solve for x by just knowing the coefficients and arranging them according to some set formula?

The general equation may be written $ax^2 + bx + c = 0$, where a, b, and c can be any integers or, in fact, any rational numbers. The symbols b and c can even be zero. If b is zero, then the equation becomes $ax^2 + c = 0$; if c is zero, the equation becomes $ax^2 + bx = 0$; and if both b and c are zero, the equation becomes $ax^2 = 0$. All these forms are still second-degree equations.

However, a cannot be allowed to equal zero, for that would convert the equation to the form $bx + c = 0$. In a general equation of any degree, the coefficient of the term to the highest power must not equal zero or the equation is reduced in degree. Even in the first-degree equation $ax + b = 0$, b may be set equal to zero, but a must not. In the latter case, the equation would become simply $b = 0$, which is no longer a first-degree equation.

The difficulty of solving the general equation of the second degree rests with the fact that it cannot be factored. What we must do then is to convert it into a form that can be factored.

First, the rules I gave for factoring in the previous chapter always involved a quadratic equation containing a simple x^2. Let's see if we can't arrange that much to begin with. To do that, let's divide both sides of the equation by a, thus:

$$\frac{(ax^2 + bx + c)}{a} = \frac{0}{a}$$

As far as the left-hand side of the equation is concerned, we know from ordinary arithmetic that $\frac{(8 + 4)}{4}$ can be written $\frac{8}{4} + \frac{4}{4}$. (Try it and see if the answer isn't 3 in both cases, and if, in other cases of the same sort, you don't get the same answer either way.)

We can therefore write the equation, thus:

$$\frac{ax^2}{a} + \frac{bx}{a} + \frac{c}{a} = \frac{0}{a}$$

In the fraction $\frac{ax^2}{a}$, we can cancel the a's. The fraction $\frac{bx}{a}$ can be written $\left(\frac{b}{a}\right)x$ and, as for the right-hand side of the equation, $\frac{0}{a}$ is, of course, 0.

Now the equation can be written:

$$x^2 + \left(\frac{b}{a}\right) x + \frac{c}{a} = 0$$

We have the simple x^2 in this form of the general equation of the second degree, but we still can't factor it. We need two quantities which when added will give the coefficient of the x term $\left(\text{which is } \frac{b}{a}\right)$ and when multiplied will give the numerical term $\left(\text{which is } \frac{c}{a}\right)$ according to the rules I gave in the previous chapter. There is no obvious way in which this can be done, however.

Why not, then, remove the $\frac{c}{a}$ term and substitute something else which will be easier to handle. To remove the $\frac{c}{a}$ term is easy. We need only transpose, thus:

$$x^2 + \left(\frac{b}{a}\right) x = -\frac{c}{a}$$

The question then arises, What do we substitute for it? Well, the simplest way of finding two values that will add up to a given quantity is to take half the quantity and add it to itself. In other words, 4 is equal to 2 plus 2, 76 to 38 plus 38, and so on. Since half of $\frac{b}{a}$ is $\left(\frac{1}{2}\right)\left(\frac{b}{a}\right)$ or $\frac{b}{2a}$, then $\frac{b}{2a}$ plus $\frac{b}{2a}$ is $\frac{b}{a}$.

So we have two values, $\dfrac{b}{2a}$ and $\dfrac{b}{2a}$, which when added give the coefficient of the x term. What will those same two values give when multiplied? The answer is:

$$\left(\frac{b}{2a}\right)\left(\frac{b}{2a}\right) = \frac{b^2}{4a^2}$$

We have to add that to the left-hand side of the equation and, in order to do that, we have to add it to the right-hand side also, so that the equation becomes

$$x^2 + \left(\frac{b}{a}\right)x + \frac{b^2}{4a^2} = -\frac{c}{a} + \frac{b^2}{4a^2}$$

Now, for just a moment, let's concentrate on the right-hand side of the equation. Suppose we multiply the fraction $-\dfrac{c}{a}$, top and bottom, by the quantity $4a$. That gives us $-\dfrac{4ac}{4a^2}$, which doesn't change the value of the fraction, but which gives it the same denominator as the second fraction on the right-hand side. The right-hand side of the equation now becomes

$$-\frac{4ac}{4a^2} + \frac{b^2}{4a^2}$$

or

$$\frac{(b^2 - 4ac)}{4a^2}$$

Our general equation can therefore be written

$$x^2 + \left(\frac{b}{a}\right)x + \frac{b^2}{4a^2} = \frac{(b^2 - 4ac)}{4a^2}$$

It is time, now, to look at the left-hand side of the equation. We have arranged it in such a way as to have two values, $\frac{b}{2a}$ and $\frac{b}{2a}$, which when added give $\frac{b}{a}$, the coefficient of the x term, and when multiplied give $\frac{b^2}{4a^2}$, which represents the term without an unknown. This means, according to the rules of the previous chapter, that the left-hand side of the equation can be factored as $\left(x + \frac{b}{2a}\right)\left(x + \frac{b}{2a}\right)$, or $\left(x + \frac{b}{2a}\right)^2$. Now the equation becomes

$$\left(x + \frac{b}{2a}\right)^2 = \frac{(b^2 - 4ac)}{4a^2}$$

We can transpose the exponent, and that will give us

$$x + \frac{b}{2a} = \pm\sqrt{\frac{(b^2 - 4ac)}{4a^2}}$$

As you see, forming the square root makes it necessary to insert a plus-or-minus sign.

As for the right-hand side, $\pm\sqrt{\frac{(b^2 - 4ac)}{4a^2}}$ can be written as $\pm \dfrac{\sqrt{b^2 - 4ac}}{\sqrt{4a^2}}$. The square root of

the numerator can't be worked out, but the square root of the denominator can be, since $(2a)(2a)$ equals $4a^2$. Consequently $\sqrt{4a^2}$ is equal to $2a$.

The equation can now be written

$$x + \frac{b}{2a} = \pm \frac{\sqrt{b^2 - 4ac}}{2a}$$

By transposition we have

$$x = -\frac{b}{2a} \pm \frac{\sqrt{b^2 - 4ac}}{2a}$$

Since the two fractions have the same denominator, we can combine them and have the equation read, at last:

$$x = \frac{-b \pm \sqrt{b^2 - 4ac}}{2a}$$

This is the general solution for x in any second-degree equation, expressed in terms of the coefficients undergoing various algebraic operations. To solve any quadratic equation, it is only necessary to put it into its general form and substitute the coefficients into the above formula for x.

In the equation $17x^2 - 2x - 5 = 0$, for instance, $a = 17$, $b = -2$, and $c = -5$. Let's substitute

these numerical values into the general solution
and we have:

$$x = \frac{-(-2) \pm \sqrt{(-2)(-2) - (4)(17)(-5)}}{(2)(17)}$$

$$x = \frac{2 \pm \sqrt{4 + 340}}{34}$$

$$x = \frac{2 \pm \sqrt{344}}{34} = \frac{2 \pm \sqrt{(4)(86)}}{34} = \frac{2 \pm 2\sqrt{86}}{34}$$

$$x = \frac{(2)(1 \pm \sqrt{86})}{(2)(17)} = \frac{1 \pm \sqrt{86}}{17}$$

Notice first that there are two answers since
there is a plus-or-minus sign involved. It is one
answer if the plus is used and a second if the minus
is used:

$$x = \frac{(1 + \sqrt{86})}{17} \quad \text{and} \quad x = \frac{(1 - \sqrt{86})}{17}$$

Notice also that $\sqrt{86}$ is an irrational number.
You see, then, that it is possible, in equations of
the second degree, to obtain a value of x that is
irrational, even though the coefficients of the equation are rational. In practical problems involving
such solutions, an approximate answer can be
found by taking the value of the irrational number
to as many decimal places as necessary. (Fortu-

nately, such values can be worked out, or even just looked up in tables.)

The value of $\sqrt{86}$, for instance, is approximately 9.32576. The two answers, therefore, are approximately 0.6074 and −0.4309.

Of course, it is possible to have rational solutions to a quadratic equation also.

In the equation $x^2 + 5x + 6 = 0$, a is equal to 1, b to 5, and c to 6. Substitute these values in the general formula and you have:

$$x = \frac{-5 \pm \sqrt{(5)(5) - (4)(1)(6)}}{(2)(1)}$$

$$x = \frac{-5 \pm \sqrt{25 - 24}}{2}$$

$$x = \frac{-5 \pm \sqrt{1}}{2}$$

Here, you see, the square root disappears, for $\sqrt{1}$ is equal to 1. The two solutions are therefore:

$$x = \frac{(-5 + 1)}{2} = \frac{-4}{2} = -2$$

or

$$x = \frac{(-5 - 1)}{2} = \frac{-6}{2} = -3$$

which, to be sure, are solutions we might have gotten directly by factoring.

HIGHER DEGREES

General solutions for equations of the first and second degree were known before the rise of algebra in the 1500's. At that time, therefore, mathematicians interested themselves in the possibility of a solution for the general equation of the third degree in terms of algebraic manipulations of the coefficients.

I won't go into the nature of the general solution, but it was discovered, and the discovery involves a certain well-known story.

It was in 1530, that an Italian mathematician named Nicolo Fontana succeeded, at last, in discovering the general solution. (Fontana had a speech imperfection and he received the nickname Tartaglia (tahr-TAH-lyuh) — the Italian word for "stammerer" — as a result. The use of the nickname was so widespread that today he is hardly ever referred to as anything but Nicolo Tartaglia.)

In those days, mathematicians sometimes kept their discoveries secret, much as industries today may keep their production methods secret. Tartaglia won great fame by being able to solve problems which involved cubic equations and which no one else could solve. Undoubtedly, he enjoyed his position as a mathematical wonder-worker.

Other mathematicians naturally kept begging

Tartaglia to reveal the secret. Finally, Tartaglia succumbed to the pressure and, in 1545, revealed the solution to Geronimo Cardano, the mathematician who introduced negative numbers and imaginary numbers. He insisted that Cardano swear to keep the matter secret.

Once Cardano had the solution, however, he promptly published it and said it was his own. Poor Tartaglia had to begin a long fight to keep the credit for himself and, ever since, mathematicians have been arguing as to who should get credit for the discovery.

Nowadays, you see, we consider it quite wrong for any scientist to keep a discovery secret. We feel he must publish it and let all other scientists know about it; that only so can science and knowledge progress. In fact, the scientific world gives credit for the discovery of any fact or phenomenon or theory to the man who first publishes it. If someone else makes the same discovery earlier but keeps it secret, he loses the credit.

According to this way of thinking, Tartaglia was wrong to keep his solution a secret and Cardano was right to publish, and rightfully deserves the credit. However, in the 1500's this was not the common viewpoint, and we should make allowance for the fact that it wasn't considered wrong at the time to keep scientific secrets.

Besides, even if Cardano was right, as a mathematician, to publish the solution, he was wrong, as a human being, to claim it was his own and not to give Tartaglia credit for thinking of it first. However, as it happens, although Cardano was a great mathematician, he was also a great scoundrel in many ways.

At about the same time, Cardano tried to work out a solution for the general equation of the fourth degree. He couldn't manage that and passed the problem on to a young man named Ludovico Ferrari, who was a student of his. Ferrari promptly solved it.

Naturally mathematicians felt that, after that, only patience and hard work were necessary to work out the solution for the general equation of any degree, but when they took up the general equation of the fifth degree, they found themselves in trouble. Nothing seemed to work. For nearly three hundred years, they tried everything they could think of and for nearly three hundred years they failed. Even Euler (the man who first used i for the square root of minus one), one of the greatest mathematicians of all time, tackled the fifth degree and failed.

Then, in 1824, a young Norwegian mathematician, Niels Henrik Abel (who was only 22 at the time, and who was to die only five years later), was able to prove that the general equation of the fifth degree was insoluble: It could not be solved in terms of its coefficients by means of algebraic operations. There were other ways of doing it, but not by algebra.

It turned out that no equation of degree higher than the fourth could be solved in this way. In 1846, a brilliant young French mathematician, Evariste Galois (ay-vah-REEST ga-LWAH), who was tragically killed in a duel at the age of only 21, found a new and more advanced mathematics, the "theory of groups," that could handle equations of high degree, but that is not for this book, of course.

BEYOND THE DEGREES

If we take a general equation of any degree, then any value which can serve as a solution for x is called an "algebraic number."

For instance, I have already said that for the general equation of the first degree any rational number, positive or negative, can serve as a solution. Even zero can serve as a solution for x, in the equation $ax = 0$. All rational numbers are therefore algebraic numbers.

For the general equation of the second degree, any rational number can serve as a solution. So also can certain irrational numbers. Thus, the square root of any rational number can serve as a solution.

For the general equation of the third degree, rational numbers, square roots, and cube roots will serve as solutions. The fourth degree will add fourth roots to the list, the fifth degree will add fifth roots, and so on.

In the end, the list of algebraic numbers includes all rational numbers and all irrational numbers that are roots (in any degree) of rational numbers.

But does this include all numbers? Are there irrational numbers which are not the roots, in one degree or another, of some rational number?

In 1844, a French mathematician named Joseph Liouville (lyoo-VEEL) was able to show that such irrational numbers did exist, but he wasn't able to show that some particular number was an example. It wasn't until 1873 that another French mathematician, Charles Hermite (ehr-MEET), turned the trick.

He showed that a certain quantity, much used in higher mathematics and usually symbolized as e, was an irrational number that was not the root, in any degree, of any rational number. (Hermite also solved the general equation of the fifth degree by nonalgebraic methods.)

The approximate value of e is 2.7182818284590-

452353602874 . . . Modern computers have worked out the value to 60,000 places. The quantity *e* was the first nonalgebraic number to be discovered and is an example of a "transcendental number" (from Latin words meaning to "climb beyond," because these existed beyond the long list of algebraic numbers).*

Another interesting quantity is the one usually represented by mathematicians as π, which is the Greek letter "pi." This quantity represents the ratio of the circumference of a circle to its diameter. If the length of the diameter of a circle is multiplied by π, the length of the circumference is found. The approximate value of π is 3.1415926535-89793238462643383279502884197169399375 10 . . . and modern computers have worked out its value to ten thousand places.

In 1882, the German mathematician Ferdinand Lindemann, using Hermite's methods, proved that π was transcendental. It is now known that almost all logarithms are transcendental; that almost all "trigonometric functions" such as the sine of an

---

* The importance of *e* rests in the fact that it is essential in the calculation of two sets of values of great importance in mathematical computations and relationships. These are logarithms (see *Realm of Numbers*) and the ratios of the sides of right triangles ("trigonometric functions"). Almost all logarithms and trigonometric functions are irrational, and those that are, are also transcendental.

angle (matters taken up in that branch of mathematics called "trigonometry") are transcendental;
that any number raised to an irrational power, such
as $2^{\sqrt{2}}$, is transcendental.

In fact, there are far more transcendental numbers than there are algebraic numbers. Although
there are more algebraic numbers than anyone can
possibly count, it remains true, even so, that almost
all numbers are transcendental.*

---

* One way of putting this is that while the set of all
algebraic numbers is infinite, the set can be represented by the lowest transfinite number. The set of
all transcendental numbers is also infinite, but can be
represented by a higher transfinite number. If you
are curious about this, I go into some detail in this
matter in the last chapter of *Realm of Numbers*.

12

Two at Once

EQUATIONS WITHOUT SOLUTIONS

So FAR, we have never considered a problem or equation in which more than one quantity was unknown. And yet it is possible to have more than one unknown.

Here's a case. Suppose you are told that the perimeter of a certain rectangle is equal to 200 inches (the perimeter being the sum of the lengths of all four sides). The question is: What are the lengths of the four sides?

To begin with, let's place the length of one side of the rectangle equal to x. The side opposite to that must also equal x (for it is one of the properties of the rectangle that opposite sides are equal in length). Together these two opposite sides are $2x$ in length.

Now what about the other pair of opposite sides? Has one any idea of what their length is?

I'm afraid not, at least not in actual numerical values. They are as unknown as the first pair and must also be given a literal symbol representing an

unknown. It would be confusing to use x because
that is already in use in this problem. It is customary, however, to use y as a second unknown.
The other pair of sides can therefore each be set
equal to y in length, giving a total of $2y$.

Now we can say that

$$2x + 2y = 200$$

We can try to solve this equation for x and hope for
the best, and we can begin by factoring:

$$2(x + y) = 2(100)$$

If we divide each side of the equation by 2, then

$$x + y = 100$$

Then, by transposing,

$$x = 100 - y$$

There's our value of x, but what good is it?
Since we don't know the value of y, we can't convert our value for x into a numerical value. Of
course, if we knew that y was equal to 1 inch, then
x would be equal to $100 - 1$ or 99 inches. Of if we
had some way of telling that y was equal to 7 inches,
x would be equal to $100 - 7$ or 93 inches. Or if
y were equal to 84.329 inches, then x would be
equal to $100 - 84.329$ or 15.671 inches.

Each pair of values would satisfy the equation.

$$2(99) + 2(1) = 200$$

$$2(93) + 2(7) = 200$$

$$2(15.671) + 2(84.329) = 200$$

You could make up any number of other pairs that would satisfy the equation, too. This doesn't mean, of course, that any two numbers at all would do. Once you pick a value for y, there is then only one value possible for x. Or if you start by picking a value for x, only one value remains possible for y.

Another warning, too. You can't pick a value for either x or y that is 100 or over without having certain practical difficulties. If you let y equal 100, then x equals $100 - 100$ or 0. You don't have a rectangle at all, then, but just a straight line. Or if you decide to let y equal 200, then x equals $100 - 200$ or -100 and you have to decide what you mean by a rectangle with a side equal to a negative number in length.

Despite practical difficulties, however, such sets of values do satisfy the equation mathematically:

$$2(0) + 2(100) = 200$$

$$2(-100) + 2(200) = 200$$

But then even if you decide to limit the values of x and y to the range of numbers greater than 0 and less than 100, there are still an endless number of pairs that you can choose which would satisfy the equation. And there would be no reason for you ever to think that one pair of values was any more correct as a solution than any other pair. You could not possibly pick among those endless numbers of pairs and for that reason an equation such as $2x + 2y = 200$ is called an "indeterminate equation."

INTEGERS ONLY

You might think that an equation without a definite solution would be dull indeed and that mathematicians would turn away from it with nose in air. Not so. Actually, such equations have fascinated mathematicians greatly.

One of the first to interest himself in such indeterminate equations was a Greek mathematician called Diophantus, who lived about A.D. 275 in the city of Alexandria, Egypt. He was particularly interested in equations where the solutions could only be integers. Here is an example of a problem leading to such an equation.

Suppose there are 8 students in a class, some boys and some girls. How many boys are there and how many girls? If we set the number of boys equal to x and the number of girls to y, then we have the equation

$$x + y = 8$$

Now we can only let x and y equal certain values. We can't have either x or y equal to $\sqrt{2}$ or to $1\frac{1}{2}$, because we can't have an irrational quantity of children or even a fractional quantity. Furthermore, we can't have either x or y equal to zero, because we have said that there are both boys and girls in the class. Nor can we let either be equal to 8, since then the other would be equal to 0, or to more than 8, for then the other would be equal to some negative number, and we don't want a negative number of children either.

For this reason, there are only a very limited number of possible solutions to the equation. If $x = 1$, then $y = 7$; if $x = 2$, then $y = 6$, and so on. In fact, there are only 7 sets of possible answers: 1 boy and 7 girls, 2 boys and 6 girls, 3 boys and 5 girls, 4 boys and 4 girls, 5 boys and 3 girls, 6 boys and 2 girls, and 7 boys and 1 girl.

Even though the number of answers is limited, the equation is still indeterminate because we have no way of telling which of the seven pairs of numbers is the correct answer, because there is no one correct answer. All are equally correct.

An indeterminate equation to which the solutions must be expressed as whole numbers only is called a "Diophantine equation" in honor of the old Greek mathematician.

Some Diophantine equations have great fame in the history of mathematics.

For instance, the Greeks were very much interested in the right triangle (a three-sided figure in which one of the angles is a right angle). A Greek mathematician named Pythagoras, who lived about 530 B.C., was able to show that the sum of the squares of the lengths of the two sides making up the right angle of the right triangle was equal to the square of the length of the side opposite the right angle (called the "hypotenuse"). In his honor, this mathematical fact is often called the "Pythagorean theorem."

We can express this algebraically, by letting the length of one side equal x, of the second equal y, and of the hypotenuse equal (what else?) z. The equation becomes

$$x^2 + y^2 = z^2$$

Since we have three unknowns in a single equation, we have an infinite number of possible solutions. Pick any values you choose for any two of the unknowns and you can work out a value for the third. If you decide to let both x and y equal 1, then z^2 equals 1^2 plus 1^2 or 2, and z is consequently equal to $\sqrt{2}$. Or if you decide to let x equal 2 and y equal 13, then z^2 equals 2^2 plus 13^2 or 173, and z equals $\sqrt{173}$.

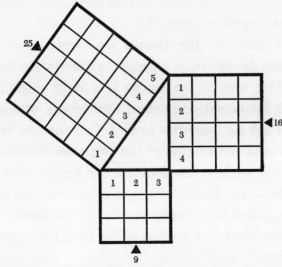

PYTHAGOREAN THEOREM

Here are two sets of three numbers, then — 1, 1, $\sqrt{2}$; and 2, 13, $\sqrt{173}$ — that satisfy the Pythagorean equation. You can find any number of additional sets with hardly any effort.

But suppose you are interested only in solutions where x, y, and z are all whole numbers, so that the equation becomes Diophantine.

You might wonder first if any such all-integer solutions exist. Well, suppose you set x equal to 3 and y equal to 4. Then z^2 is equal to 3^2 plus 4^2 or 25, and z is equal to $\sqrt{25}$ or 5. There you have a set of integers — 3, 4, and 5 — which serve as a solution to the equation.

There are other such solutions, too. For instance

$5^2 + 12^2 = 13^2$, so that 5, 12, and 13 are a solution.
In fact, there turn out to be an endless number of
such all-integer solutions to the equation.   Mathematicians have worked out rules for finding such
solutions and in doing so have learned a great deal
about the handling of whole numbers.

In the early 1600's, there lived a French mathematician named Pierre de Fermat (fehr-MAH),
who studied the behavior of integers so thoroughly
that he founded a branch of mathematics dealing
with integers and called "the theory of numbers."

Fermat had a habit of scrawling in the margins
of books he was reading and one time he wrote that
he had discovered an interesting fact about equations of the type

$$x^n + y^n = z^n$$

where n can equal any whole number.   (The Pythagorean equation is the special variety of this group
in which n equals 2.)

Fermat wrote that he had found that whenever
n was greater than 2 in such an equation, there
were no solutions that consisted of integers only.
In other words, you could add the squares of two
whole numbers and end with the square of another
whole number as in $3^2 + 4^2 = 5^2$, but you couldn't
add the cubes of two whole numbers and ever end
with the cube of another whole number, or add the

fourth powers of two whole numbers and ever end with the fourth power of another whole number, and so on.

Fermat wrote in the book that he had a beautiful and simple proof of this, but that the margin was too small to contain it. He never did write the proof (or, if he did, no one has ever found it) and what is called "Fermat's Last Theorem" has never been proved to this day.

But mathematicians searched for it. Fermat was such a brilliant worker that they couldn't believe he had made a mistake. Surely the proof existed. Every great mathematician had a try at it. Prizes were offered. It seems to be true — but no one has ever found the proof to this day.

Probably Fermat was mistaken in thinking he had a proof, but we can never be sure.

If only the margin of the book had been a little bigger.

ADDING TO THE INFORMATION

What makes for an indeterminate equation is lack of information. If we are told that the boys and girls in a class total 8, then all we can say is that $x + y = 8$ and there is no clear solution. But suppose our information is increased. Suppose we are also told that there are three times as many girls as boys. If we have decided to let x equal the number of boys and y the number of girls, then we can say that $y = 3x$.

So our information now enables us to set up two equations, each in two unknowns:

$$x + y = 8$$

$$y = 3x$$

These are called "simultaneous equations," because the same values of x and y must simultaneously satisfy both equations.

Since $y = 3x$, we can naturally substitute $3x$ for y wherever y occurs. In particular, we can substitute $3x$ for y in the first equation and get

$$x + 3x = 8$$

Suddenly, we have an equation with only one unknown, and a very simple equation at that, which works out to:

$$4x = 8$$

$$x = \frac{8}{4}$$

$$x = 2$$

The number of boys is 2, and we can now substitute 2 for x wherever that occurs. We can do it in the equation $y = 3x$, which becomes $y = 3(2)$ or 6. Our final solution then is that there are 2 boys and 6 girls in the class. The sum is indeed 8 and there are indeed three times as many girls as boys.

The same principle works in more complicated situations. Suppose we have two equations as follows:

$$7x - 3y = 7$$

$$5x + 2y = 34$$

Let's solve for x in the first equation:

$$7x - 3y = 7$$

$$7x = 7 + 3y$$

$$x = \frac{(7 + 3y)}{7}$$

Now we can substitute $\frac{(7 + 3y)}{7}$ for x in the second equation and get:

$$\frac{5(7 + 3y)}{7} + 2y = 34$$

$$\frac{(35 + 15y)}{7} + 2y = 34$$

$$\frac{35}{7} + \frac{15y}{7} + 2y = 34$$

$$5 + \frac{15y}{7} + 2y = 34$$

$$\frac{15y}{7} + 2y = 34 - 5$$

$$\frac{15y}{7} + 2y = 29$$

I hope the reasons for each step have been clear to you so far. Now let's remove fractions, just as we would in arithmetic by multiplying each side of the equation by 7:

$$7\left(\frac{15y}{7} + 2y\right) = 7(29)$$

$$7\left(\frac{15y}{7}\right) + 7(2y) = 7(29)$$

$$15y + 14y = 203$$

$$29y = 203$$

$$y = \frac{203}{29}$$

$$y = 7$$

Now that we know that $y = 7$, we can go back to either of the original equations and substitute 7 for y. In the first equation:

$$7x - 3(7) = 7$$

$$7x - 21 = 7$$

$$7x = 7 + 21$$

$$7x = 28$$

$$x = \frac{28}{7}$$

$$x = 4$$

Or, if we prefer to use the second equation:

$$5x + 2(7) = 34$$

$$5x + 14 = 34$$

$$5x = 34 - 14$$

$$5x = 20$$

$$x = \frac{20}{5}$$

$$x = 4$$

Either way, the sole answer we get is that x is equal to 4 and y is equal to 7, and if both values are substituted in either equation, you will find that they are valid solutions.

What's more, if we had solved for x in the second equation and substituted its value in the first, or if we had solved for y in either equation and substituted its value in the other, we would have ended with the same solution; x is 4 and y is 7. (You might try it for yourself and see.)

In fact, we could proceed by solving both equations for either x or y; let's say for y. In the first case:

$$7x - 3y = 7$$

$$-3y = 7 - 7x$$

$$3y = 7x - 7$$

$$y = \frac{(7x - 7)}{3}$$

In the second case:

$$5x + 2y = 34$$

$$2y = 34 - 5x$$

$$y = \frac{(34 - 5x)}{2}$$

We now have two different expressions for y. If we are to avoid inconsistency, we must assume that the two different expressions have the same value, so we can set them equal to each other:

$$\frac{(7x - 7)}{3} = \frac{(34 - 5x)}{2}$$

Now we have a single equation with one unknown. We can begin by clearing fractions in the usual arithmetical way of multiplying both fractions by the product of the two denominators, (3)(2) or 6:

$$\frac{6(7x - 7)}{3} = \frac{6(34 - 5x)}{2}$$

$$2(7x - 7) = 3(34 - 5x)$$

$$14x - 14 = 102 - 15x$$

$$14x + 15x = 102 + 14$$

$$29x = 116$$

$$x = \frac{116}{29}$$

$$x = 4$$

And, of course, if we substitute 4 for x in either of the original equations, we find that y turns out to be equal to 7.

AND STILL ANOTHER WAY

There is still one more device we can apply to our two unknowns in two equations. To understand this new device, let's begin by considering two very simple general equations: $a = b$ and $c = d$.

We know that a particular value can be added to both sides of an equation or subtracted from both sides without making the equation false. So far

I have always added or subtracted the same expression on both sides, but that is really not necessary. I can add (or subtract) different expressions provided they have the same value. In other words I can add 5 to one side of the equation and $17 - 12$ to the other side.

Well, then, if $c = d$, I can add c to one side of an equation and d to the other without making the equation false. To put it in symbols:

If $a = b$

and $c = d$

then $a + c = b + d$

or $a - c = b - d$

or $ac = bd$

or $\dfrac{a}{c} = \dfrac{b}{d}$

and so on.

Now we can go back to our two equations:

$$7x - 3y = 7$$

$$5x + 2y = 34$$

By the general rule I have just discussed, I can add the left-hand side of the second equation to the left-hand side of the first, and the right-hand side of the second equation to the right-hand side of the first. I get:

$$(7x - 3y) + (5x + 2y) = 7 + 34$$

$$7x + 5x - 3y + 2y = 41$$

$$12x - y = 41$$

You may well ask what I have accomplished and the answer is nothing. But wouldn't it have been nice if in adding I could have gotten rid of either x or y. If we could arrange the y term in one equation to equal 0 when added to the y term in the other, we could do just that. And here's how.

Suppose I multiply the first equation by 2, both right and left, as I can without spoiling the equation. I get:

$$2(7x - 3y) = 2(7)$$

$$14x - 6y = 14$$

Then, suppose I multiply the second equation, left and right, by 3:

$$3(5x + 2y) = 3(34)$$

$$15x + 6y = 102$$

As you see by this arrangement I have managed to have a $-6y$ in one equation and a $+6y$ in the other. Now, if I add the two equations, these two

terms add up to zero and there is no longer a y term
in the equation:

$$(14x - 6y) + (15x + 6y) = 14 + 102$$

$$14x + 15x - 6y + 6y = 116$$

$$29x = 116$$

$$x = \frac{116}{29}$$

$$x = 4$$

and, by substitution, y will equal 7 again.

By a number of different devices, then, it is always possible to take two equations, each containing two unknowns, and make of them one equation
containing one unknown. One important point to
remember, however, is that the second equation
must be independent of the first; that is, it must
really add new information.

If one equation can be converted into the second
by adding the same value to both sides, or by subtracting, multiplying, dividing, raising to a power,
or taking a root on both sides equally, they are
really the same equation. No new information is
added by the second.

To take a simple case, suppose you had:

$$x - y = 2$$

$$2x - 2y = 4$$

You could convert the first equation to the second by simply multiplying each side by 2; or you could convert the second equation to the first by dividing left and right by 2. They are therefore the same equation. If you ignore that and decide to go ahead anyway and see what happens, you can solve the first equation for x and find that $x = 2 + y$.

Next substitute $2 + y$ for x in the second equation:

$$2(2 + y) - 2y = 4$$

$$4 + 2y - 2y = 4$$

The terms containing y add up to zero and you have left only that

$$4 = 4$$

which is certainly true but doesn't help you much in determining the value of x and y. This is an example of "arguing in a circle."

Naturally, if you have three unknowns, you need three independent equations. Equations 1 and 2 can be combined to eliminate one of the unknowns and equations 1 and 3 (or 2 and 3) can be combined to eliminate that same unknown. That gives you two equations with two unknowns, from which point you can proceed in the manner I have just given you.

In the same way, numerical values for four unknowns can be found if four independent equations exist; five unknowns if five independent equations exist, and so on. The process quickly gets tedious, to be sure, and special techniques must be used, but the mere number of unknowns should never be frightening in principle — so long as you have enough information to work with.

13

Putting Algebra to Work

GALILEO ROLLS BALLS

IT MAY BE that, as you read this book, the thought occurs to you: But what good is all this?

I'm sure you know in your heart that mathematics is really very useful, but as you try to follow all the ways in which equations must be dealt with, you may still get a little impatient. Is algebra really worth all the trouble it takes to learn it?

Of course, it can come in handy in solving problems that come up in everyday life. For instance, suppose you have $10 to spend but intend to shop at a store where all sales are at 15 per cent discount from the list price. If you have a catalog giving you only the list prices, what is the cost of the most expensive item you can buy?

You might, if you wished, just take some prices at random and subtract 15 per cent until you found one that gave you a discount price of $10. That would be clumsy, however. Why not tell yourself instead that with a 15 per cent discount, you are paying 85 per cent or $\frac{85}{100}$ of the list price, so that $\dfrac{85}{100}$ of some unknown value which you will call x is \$10. The equation is:

$$\frac{85x}{100} = 10$$

and you ought to be able to solve it easily according to the principles discussed in this book:

$$x = \frac{(10)(100)}{85} = 11\frac{65}{85}$$

To the nearest penny, this comes out \$11.77. If you check this, you will find that a 15 per cent discount of \$11.77 is, to the nearest penny, \$1.77, leaving you a net price of \$10.00.

Or suppose someone is following a recipe which is designed to make 4 helpings of a particular dish, while what is needed is 7 helpings. The cook will naturally want to increase all quantities of ingredients in proportion. In real life, this is usually done by guess, which is why cooking sometimes turns out badly even though a recipe is before your eyes.

Why not use algebra? If you are required to add $1\frac{1}{2}$ teaspoons of flavoring in the original recipe, you can say, "Four is to seven as one and a half is to something I don't know yet." One way of writing this statement mathematically is: $4:7::1.5:x$, which puts it in the form of "ratios."

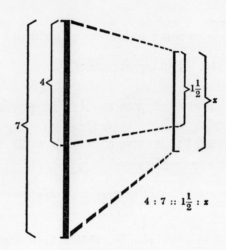

$$4 : 7 :: 1\frac{1}{2} : x$$

However, when I mentioned fractions on page 41, I said they could be considered as ratios, so the equation can also be written as

$$\frac{4}{7} = \frac{1.5}{x}$$

and the equation can be solved as follows:

$$4x = (1.5)(7) = 10.5$$

$$x = \frac{10.5}{4} = 2.625$$

Naturally, you are not going to add flavoring to the nearest thousandth of a teaspoon, but you can try to add just a trifle over $2\frac{1}{2}$ teaspoons, which will be far better than just making a wild guess.

Of course, these are little things and you might feel that algebra isn't very important if it is used only in calculating prices and adjusting recipes.

But that is *not* all it is used for. The main use of algebra arises in connection with the attempts of scientists to understand the universe. Let's see how it can be used in that connection, and how a few symbols can help achieve the most astonishing accomplishments. I'll begin at the beginning.

The ancient Greek philosophers were very interested in the shapes and forms of objects, so they developed geometry to great heights. They were not interested in actually measuring and weighing and so they did not develop algebra.

The result was that their notions were a little fuzzy. For instance, the greatest of the Greek philosophers, Aristotle, was interested in the way bodies moved when left to themselves. He was content, however, to say that solid and liquid bodies moved downward, spontaneously, toward the center of the earth, while air and fire moved, spontaneously, upward away from the center of the earth. Moreover, he stated that heavy objects moved downward (or fell) more quickly than light objects, so that a stone fell more rapidly than a leaf, for instance.

He did not think to try to measure how quickly a stone fell, or whether it fell at different speeds at the beginning of its fall as opposed to the end of its fall.

Toward the dawn of modern times, things began to change. The Italian artist Leonardo da Vinci, near the end of the 1400's, suspected that falling objects increased their rate of movement as they fell. But it wasn't till a century later, toward the end of the 1500's, that the Italian scientist Galileo Galilei (usually known by his first name only) actually set out to measure the rate of fall.

To do this wasn't easy, because Galileo had no clocks to work with. To keep time, he had to use his pulse, or he had to measure the weight of water pouring out of a small hole at the bottom of a water-filled bucket. This wasn't good enough to measure the short time-intervals involved in studying objects falling freely as a result of the action of gravity. What he did then was to roll balls down gently sloping tracks.

In this way, balls rolled more slowly than they would if they were falling freely. The pull of gravity was, so to speak, diluted, and Galileo's crude time measurements were good enough.

In such experiments, Galileo found that the velocity with which a ball moved down an inclined plane was directly proportional to the time it moved. (I won't bother describing how he measured velocities, since that is not what we are interested in here.)

Thus, it might be that the ball, after starting from a standing position at the top of the tracks, would be moving 3 feet per second at the end of one second. At the end of twice the time (2 seconds), it would be moving at twice the velocity (6 feet per second). At the end of four times the time (4 seconds), it would be moving at 4 times the velocity (12 feet per second), and so on.

As you see, in order to work out the velocity, it is only necessary to multiply the time during which the ball has been rolling by some fixed number. In the case I have just described, the fixed number or "constant" is 3, so that you can decide at once that after 37.5 seconds the ball would be moving at the velocity of (37.5)(3) or 112.5 feet per second.

To express this generally, we can let the time during which the ball has been rolling be symbolized as t. (It is often customary in experiments of this sort to symbolize different quantities by initial letters.) The velocity, therefore, is symbolized as v and the constant as k (which has the sound of the initial letter c, anyway).

The equation Galileo could write to represent his discovery about moving bodies was

$$v = kt$$

Now k had a constant value in one particular experiment, but Galileo found that this value would shift if he changed the slant of his tracks. As the track was made to slant more gently, the value of k declined, and if the track was made steeper, its value increased.

Clearly, the value of k would be highest if the track was as steep as it could get — if it were perfectly vertical. The ball would then be falling freely under the pull of gravity, and the constant could then be symbolized as g (for gravity). The equation of motion for a freely falling body is, then,

$$v = gt$$

From the experiments with inclined planes, Galileo could calculate the value of g (not by ordinary algebra, to be sure, but by another branch of mathematics called trigonometry) and this turned out to be equal to 32, so that the equation becomes

$$v = 32t$$

This means that if a body is held motionless above the surface of the earth and is then dropped and allowed to fall freely, it would be moving at the end of one second at the rate of 32 feet per second; at the end of two seconds, 64 feet per second; at the end of three seconds, 96 feet per second; and so on.

Galileo also found by experiment that the same equation of motion held for all bodies rolling or falling downward under the pull of gravity — all bodies, however heavy or light. There was a second force, the resistance of air, which counteracted the pull of gravity, but which was very weak so that it showed a noticeable effect only on very light bodies that offered a large surface to the air — feathers, leaves, pieces of paper, and so on. These fell slowly and that was what had deluded Aristotle into thinking the force of gravity was different on different bodies.

Galileo also measured the distance covered by a body rolling down an inclined plane. Naturally, since its velocity was increasing, it covered more ground each second than it did the second before. In fact, Galileo found by experiment that the total distance was directly proportional to the square of the time. In 3 seconds it covered (3)(3) or 9 times the distance it covered in 1 second. In 17 seconds, it covered (17)(17) or 289 times the distance it covered in 1 second, and so on.

The equation worked out by Galileo for the distance, d, covered by a freely falling body was

$$d = \frac{gt^2}{2}$$

Since g is equal to 32, this equation works out to

$$d = 16t^2$$

This means that, after 1 second, a freely falling

body covers a distance of (16)(1)(1) or 16 feet; after 2 seconds, it covers a distance of (16)(2)(2) or 64 feet; after 7 seconds, a distance of (16)(7)(7) or 784 feet, and so on.

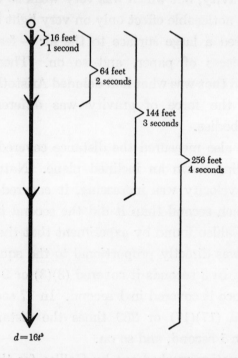

$d = 16t^2$

In this way, Galileo was able to express the behavior of moving bodies by means of algebraic equations. This meant that the behavior could be described in sharper, clearer fashion than by words alone. Furthermore, by making use of equations, problems involving falling bodies could be solved by making use of the algebraic techniques that mathematicians had been working out during the previous century.

The study of moving bodies moved ahead swiftly as a result and the whole world of scholarship was treated to the spectacle of how knowledge increased once mathematics and mathematical techniques were applied to natural phenomena. As a result, algebra (and, eventually, higher mathematics, too) came to seem essential to science and, in fact, the birth of modern science is dated with Galileo's experiments on rolling bodies.

NEWTON DEDUCES GRAVITATION

The equations used in expressing experimental observations can be used to deduce important generalizations about the universe.

For instance, suppose Galileo's inclined plane were made to slope as gently as possible; in other words, suppose it were to be perfectly horizontal. In that case, k would equal 0. (You could find this out by direct experiment with a horizontal plane, or calculate it, by means of trigonometry, from experiments with planes that are not horizontal.) The equation of motion on a horizontal surface is, therefore:

$$v = 0t$$

or

$$v = 0$$

This means that a ball resting on a horizontal surface remains motionless.

Now suppose that a ball were moving at a fixed velocity, which we can represent as V, and were then to start rolling down an inclined plane. Its velocity would increase according to the equation we had already used, but at every point, there would be added, to that changing velocity, the fixed velocity with which it had started. In other words,

$$v = kt + V$$

But suppose the inclined plane were horizontal so that k equaled zero. The motion of a ball that had started with a fixed velocity would then be, according to the equation,

$$v = 0t + V$$

or

$$v = V$$

In other words, a body moving at a fixed velocity under conditions in which gravity or some other force could not act upon it, would continue to move at that fixed velocity. There is no term involving t in the equation so there is no change in velocity with time.

The scientist who expressed this clearly for the first time was the English mathematician Isaac Newton, who was born in 1642, the year Galileo died.

Newton said that every object at rest remains
at rest unless acted on by an outside force such as
gravity, and every body in motion continues to
move at a constant velocity in a straight line unless
acted on by an outside force such as gravity.

This is the "First Law of Motion" or the "principle of inertia."

None of the ancient philosophers had stumbled
on this truth. They thought that a body in motion
tended to come to rest spontaneously, unless some
continuing force kept it in motion. The reason they
thought this was that actual phenomena are complicated. Rolling balls seem to come spontaneously
to rest, if set rolling on a level surface, because of
the action of outside forces such as air resistance
and friction.

Even Newton, perhaps the greatest thinker of all
time, might not have seen the First Law of Motion
if he could do no more than watch the behavior of
objects actually moving on the surface of the earth.
His principle arose out of a consideration of Galileo's
equations, which were deliberately simplified by
ignoring the action of air resistance and friction.

It has often proven to be the case since Newton's
time, too, that the use of equations has succeeded
in simplifying natural phenomena to the point where
an underlying pattern could be seen.

Newton worked out two other laws of motion,

in similar fashion, and these are called the "Three Laws of Motion."

In the century between 1550 and 1650, great new astronomical discoveries had been made. The Polish astronomer Nicolaus Copernicus maintained that the sun was the center of the solar system, and that the earth was not. (Most of the old Greek philosophers, including Aristotle, had insisted the earth was central.) Then, the German astronomer Johann Kepler showed that the planets, including Earth, moved about the sun in ellipses and not in circles, as previous astronomers had thought.

The question then arose as to just why planets should be moving about the sun in ellipses and at varying velocities according to their distance from the sun. (Kepler also worked out what these velocities must be.) Both Kepler and Galileo felt there must be some force attracting the planets to the sun, but neither was quite able to make out just how that force worked.

Newton realized that, in outer space, there was nothing to create friction or resistance as planets moved, and that the Laws of Motion would therefore work perfectly. He manipulated the equations representing those laws in such a way as to show that the force between any two bodies in the universe was directly proportional to the amount of matter (or "mass") in one body multiplied by the mass of the other.

Furthermore, this force as it moved away from a particular body could be imagined as stretching out in a gigantic sphere that grew continually larger and larger. The force would have to stretch out over the surface of that sphere and get weaker as it had more area to cover.

How did the area of a sphere vary with its size?

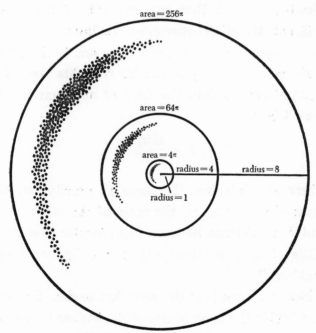

multiplying radius by 4 multiplies area by 16
multiplying radius by 8 multiplies area by 64

Well, the area of a sphere (A) varied according to the square of the radius (r), which is the distance from the center to the surface of the sphere. The exact formula is

$$A = 4\pi r^2$$

where π is the quantity I referred to at the end of Chapter 11.

Since the area of the sphere increased according to the square of the distance from center to surface, Newton decided that the strength of the gravitational attraction between two bodies must weaken as the square of the distance from one to the other.

He was now ready to put his thoughts about the force of gravity into the form of an equation and here it is:

$$F = \frac{Gm_1m_2}{d^2}$$

where F symbolizes the force of gravity, m_1 the mass of one body, m_2 the mass of the other body, and d the distance between them (center to center), while G is a constant called the "gravitational constant."

Now let's see how the equation works. Suppose you double the mass of one of the bodies; instead of m_1 you have $2m_1$. The expression $\dfrac{Gm_1m_2}{d^2}$ becomes

$\dfrac{G(2m_1)m_2}{d^2}$, which can be written $\dfrac{2Gm_1m_2}{d^2}$. The new expression is just twice as large as the old, which means that doubling the mass of one of the bodies doubles the gravitational force. If you double the mass of both bodies, you have $\dfrac{G(2m_1)(2m_2)}{d^2}$ or $\dfrac{4Gm_1m_2}{d^2}$, or four times the gravitational force.

Suppose you triple the distance between the bodies. The expression becomes $\dfrac{Gm_1m_2}{(3d)^2}$ or $\dfrac{Gm_1m_2}{9d^2}$ or $\left(\dfrac{1}{9}\right)\dfrac{Gm_1m_2}{d^2}$, showing that the force is now only $\dfrac{1}{9}$ what it was, or, in other words, has weakened ninefold. It has weakened, you see, as the square of the distance, which has increased only threefold.

CAVENDISH WEIGHS THE EARTH

Newton's equation was found to explain, quite exactly, the motion of all the bodies in the solar system. This impressed the scholars of the 1700's so much that all of them tried to imitate Newton in making great generalizations from small beginnings, and to solve all problems by reasoning. The century is, in fact, referred to as the "Age of Reason."

In the 1800's Newton's equation gained still

more fame, when it was found to apply even to distant double stars, circling each other, trillions upon trillions of miles from Earth. Then, when a newly discovered planet, Uranus, was found not quite to obey Newton's equation, astronomers deduced the existence of a still undiscovered planet whose attraction pulled Uranus out of line. That undiscovered planet was searched for and found at once — thanks to the manipulations of algebraic equations.

Let me give you an example of the sort of thing Newton's equation can do.

Suppose you were holding a stone at the lip of the Grand Canyon and let go. It would start falling. At the end of each second (ignoring air resistance) it would be falling 32 feet per second more quickly than at the end of the previous second. This increase of speed (or "acceleration") is considered, according to Newton's Second Law of Motion, to be equal to the force of Earth's gravitational pull upon the stone.

Most scientists don't like to use feet to measure length, but prefer to utilize the metric system* and to measure length in "centimeters." A centi-

---

* The metric system is discussed in considerable detail in *Realm of Measure*.

meter is equal to about $\frac{2}{5}$ of an inch and gravitational acceleration comes to 980 centimeters per second each second. Thus F, in Newton's equation, can be set equal to 980.

Next, suppose the stone we are holding weighs exactly 1 gram. (A gram is a measure of weight in the metric system and is about $\frac{1}{28}$ of an ounce.) Its distance from the center of the earth is about 3959 miles or 637,100,000 centimeters.

If we substitute 980 for F, 1 for m_1, and 637,100,-000 for d in Newton's equation, we have:

$$980 = \frac{G(1)(m_2)}{637,100,000^2}$$

$$980(637,100,000)^2 = Gm_2$$

$$Gm_2 = 398,000,000,000,000,000,000$$

Now it would be exciting if we could solve for m_2, which represents the mass of the earth, but all we can say, by transposing, is that

$$m_2 = \frac{398,000,000,000,000,000,000}{G}$$

and we don't know the value of G, which is the gravitational constant. Newton didn't know, and no one after him, for a century, knew.

In 1798, however, an English scientist named Henry Cavendish tried an experiment. This is what he did. He suspended a light rod by a wire to its center. At each end of the rod was a light lead ball. The rod could twist freely about the wire and a light force applied to the balls would produce such a twist. Cavendish calculated how large a force would produce how large a twist by actual experiment.

Now he brought two large balls near the two light balls, on opposite sides. The force of gravity between the large balls and the light ones twisted the wire. From the amount of twist, Cavendish could calculate the amount of gravitational force (F). He knew the masses of his various balls (m_1 and m_2) and the distance between them, center to center (d).

Let's take simple values just to show how things worked out, and suppose the heavy balls (m_2) weighed 1000 grams, and the light ones (m_1) weighed 1 gram; that they were at a distance (d) 10 centimeters apart, and that Cavendish calculated the gravitational force between them to be equal to 0.000000667 — a very small force, as you see.

Substituting in Newton's equation, we have

$$0.000000667 = \frac{G(1)(1000)}{(10)^2}$$

Therefore, by ordinary arithmetic,

$$0.000000667 = 10G$$

and, by transposition,

$$\frac{0.000000667}{10} = G$$

so that

$$G = 0.0000000667$$

If we return to our earlier equation for the mass of the earth

$$m_2 = \frac{398,000,000,000,000,000,000}{G}$$

and substitute 0.0000000667 for G, we have

$$m_2 = \frac{398,000,000,000,000,000,000}{0.0000000667}$$

which works out to

$$m_2 = 6,000,000,000,000,000,000,000,000,000 \quad \text{grams}$$

or

$$m_2 = 6,600,000,000,000,000,000,000 \quad \text{tons}$$

Thus Cavendish, by means of a careful experiment involving small balls, plus the techniques of algebra, had succeeded in weighing the earth.

Does this impress you with the usefulness of algebra?

If it does, look deeper. The real importance of algebra, and of mathematics in general, is not that it has enabled man to solve this problem or that, but that it has given man a new outlook on the universe.

From the time of Galileo onward, mathematics has encouraged man to look at the universe with the continual question: "Exactly how much?"

In doing so, it brought into existence the mighty structure of science, and that structure itself is far more important than any fact or group of facts that merely make up part of the structure.

Index